The Worcester Tornado: A New Perspective

by:

John L. Bisol

The Worcester Tornado: A New Perspective

ISBN: 978-1-329-70645-3

Introduction and Commentary: Copyright © 2015 by John L. Bisol

Books by Bisol

This document is a derivative of an earlier circulated report and as such this document may contain certain copyrighted material the use of which has not always been specifically authorized by the original author. This material is made available in an effort to advance the understanding of the environmental, political, economic, scientific, and disaster preparedness/response issues, etc., associated with the devastating Worcester, MA tornado of June 1953.

If you wish to use copyrighted material from this document for purposes of your own that go beyond 'fair use', you must obtain permission from the original copyright owner.

Printed in the United States of America.

10 9 8 7 6 5 4 3 2 1

Originally published in different format: 1956
Original Publisher: Washington, National Academy of sciences, National Research Council
Original Published: 1953
Original Pages: 226
Possible Copyright Status: NOT_IN_COPYRIGHT
Language: English
Original Creator: "Wallace, Anthony F. C., 1923 – [from old catalog]"

TABLE OF CONTENTS

ABSTRACT	5
INTRODUCTION: John L. Bisol	6
ABOUT: Anthony F.C. Wallace – The Original Author of the report	11
ORIGINAL TITLE PAGES/ACKNOWLEDGEMENTS	12
COMMITTEE ON DISASTER STUDIES (1953)	14
ORIGINAL TITLE PAGE:	
An Exploratory Study of Individual and Community Behavior in an Extreme Situation	15
FOREWORD (ORIGINAL)	16
PREFACE (ORIGINAL)	17
INTRODUCTION: (ORIGINAL): *A Time-Space Model of Disaster as a Type of Behavioral Event*	18
"Total Impact"	20
"Fringe Impact"	20
"Filter Area"	20
"Organized Community and Regional Aid"	21
1. STEADY STATE	22
2. WARNING	23
3. THREAT	24
4. IMPACT	25
5. ISOLATION	26
6. RESCUE	27
7. REHABILITATION	28
8. IRREVERSIBLE CHANGE	29
I: STEADY STATE: The City Before Impact	32
1. General Characteristics	32
2. Counter measure Agencies	36
II: WARNING: The City Learns Of Impending Impact	40
III: IMPACT: The Tornado Strikes	47
IV: ISOLATION: The Impact Area Goes It Alone	59
V: RESCUE: Extrication, First Aid, Reassurance, & Evacuation	69
VI: REHABILITATION: The Attempt To Restore The Steady State	81
1. Hospitalization	82
2. Welfare Services	84
VII: IRREVERSIBLE CHANGE: The City Achieves A New Equilibrium	88
VIII: SPECIAL TOPICS	91
1. The Disaster Syndrome	91
2. The Counter-Disaster Syndrome	110
3. The Length of the Isolation Period	115
4. The "Cornucopia Theory"	117
BIBLIOGRAPHY (Original)	121
LIST OF INTERVIEWS AND FIELD NOTES	123
NATIONAL ACADEMY OF SCIENCES – NATIONAL RESEARCH COUNCIL	124
ADDITIONAL RESOURCES	126
ENDNOTES – John L. Bisol	127

MAPS

1. The City of Worcester	19
2. Ecological Zones in Worcester	33
3. Distribution of Security Agencies	39
4. The Path of the Tornado	41
5. Disaster Space at I-Time (5:08 - 5:20 PM)	48
6. Neighborhoods in Impact Area	52
7. Disaster Space at 5:30 PM	65
8. Disaster Space at 6:00 PM	71
9. Disaster Space at 8:00 PM	77
10. SPECIAL MAP: THE PATH OF THE 1953 TORNADO SUPERIMPOSED ON A MAP OF WORCESTER, MASSACHUSETTS (PRESENT DAY)	126

PHOTOGRAPHIC PLATES

1. The Tornado Cloud at Two Miles Distance	43
2. Devastation at Assumption College	56
3. Devastation in St. Nicholas Development Area	57
4. Devastation in Great Brook Valley and Curtis Apartments	58
5. Rescue and First Aid in Great Brook Valley	74
6. Evacuation from Great Brook Valley	75
7. Evacuation from Burncoat Street Area	78
8. Evacuation from Great Brook Valley	80
9. The Disaster Syndrome: Silence and Immobility	92
10. The Disaster Syndrome: The "Dazed" Reaction	96
11. The Disaster Syndrome: Body Contact	98
12. The Disaster Syndrome: Passivity	102
13. The Disaster Syndrome: The "Staring" Reaction	105

TABLES

TABLE 1: TIME STAGES IN DISASTER	18
TABLE 2: DISTRIBUTION OF INTERVIEWEES: BY LOCALITY (see map)	45
TABLE 3: INTERPRETATIONS OF SIGHT OF FUNNEL	45
TABLE 4: FATAL CASUALTIES BY NEIGHBORHOOD	53
TABLE 5: FATAL CASUALTIES BY AGE AND SEX	54
TABLE 6: TYPES OF MAJOR INJURY	54
TABLE 7: THE BREAKDOWN OF HOSPITALIZATIONS	79

INTERVIEWS

A neuro-psychiatrist who saw victims at Memorial Hospital from 5:15 PM to 2:00 AM	97
The Superintendent of St. Vincent's Hospital	98
A Nursing Adviser and Coordinator	98
In an Interview at Memorial Hospital	98
The Superintendent of Worcester State Hospital	99
The Director, Hahnemann Hospital	99
The Superintendent of Worcester State Hospital	99
A Civil Defense Secretary	100

ABSTRACT

This book re-examines the work of the Committee on Disaster Studies, Study #3, "Tornado in Worcester, Massachusetts: An Exploratory Study of Individual and Community Behaviors in an Extreme Situation" originally edited by Anthony F. C. Wallace, PhD.

The review draws upon the primary source including the original interviews, maps and committee investigatory materials (included in their entirety and unabridged). Commentary is provided by Endnotes – to enhance the material and to allow for a modern day understanding and comparison to the events described in the original report and subsequent commentary on disasters/studies.

Included is a personal recount of the day of the storm and the following day's trip to the disaster city.

INTRODUCTION: John L. Bisol

Some psychologists believe that young children cannot form lasting memories. They claim "… by the age of seven, these children could still recall more than 60 percent of the recorded events, but children who were just a year older remembered only about 40 percent. Age seven seems to mark the onset of childhood amnesia…" (Berit Brogaard D.M.Sci., Ph. D, "The Superhuman Mind: Why don't We Remember our Early Childhoods?" How the brain gets rid of unused or ineffective brain connections. Posted Feb 11, 2015 Psychology Today). The reason childhood amnesia sets in around that time has to do with "pruning", (Or so say the scholars), the main purpose of which is to get rid of unused or ineffective brain connections. Pruning is a process that changes the neural structure by reducing the overall number of synapses, or brain connections. This results in more efficient synaptic configurations. Pruning is governed primarily by environmental factors, particularly learning. (All I can say is that no one asked me what I remember.) I was five years old in 1953 and I remember the tornado of 1953:

It had been a hot, humid day, but I had no way to gauge whether it was "worse" than any other. To me, at my age (5) then, every summer day was hot and humid and miserable. Oddly, the weather seemed to cool a bit, just after 3 PM. Towards five o'clock we noticed a change in the southwest sky. At first we thought the storm would be like any other and skirt to our south to be followed by another storm to directly hit us. This storm was definitely different. My father made us put the canvas covers on the greenhouse because he expected hail.

I remember the profound, deathly stillness that enveloped us that late afternoon. The chickens needed no urging to get into their coops, they waddled silently up the ramps and sat in eerie silence on their nests. The cats skulked into hiding. No birds flew or sang. The southern sky was half covered by a blackness that was strangely "squirming", yet impenetrable. From our vantage point we could not see the cloud tops, just oppressive blackness. The northern sky displayed a vibrant blue that faded to yellow as it met the storm cloud. Something massive was moving briskly east, we could sense it, but could not see it – the entire cloud mass marched along. We heard thunder, sometimes muffled, sometimes sharp but the lightning was in the distance and indistinct. "Boy, Worcester's really getting it!" my father said. Worcester "really" was "…getting it" as a sometimes F4, sometimes F5 tornado ripped street after street of houses into piles of lumber.

From every description the Worcester tornado could be called a "Wedge" – an informal storm observers' slang for a tornado which *looks* wider than the distance from ground to ambient cloud base. There is no scientific meaning to it; since many factors (actual tornado size, cloud base height, moisture content of the air, intervening terrain, soil and dust lofting) can regulate a tornado's apparent width. Although some storm observers use the term "Wedge" rather loosely. Many famous "wedge" tornadoes have also been violent, producing EF4>EF5 damage on the Fujita scale, a tornado's size does not necessarily indicate anything about its strength. (NOAA)

The storm stayed on the ground for nearly 90 minutes, traveling 48 miles across Central Massachusetts. In total, 94 people were killed, making it the 21st deadliest tornado in the history of the United States. In addition to the fatalities, over 1,000 people were injured and 4,000 buildings were damaged (some obliterated). In 1953 the tornado caused $52 million in damage, which translates to $349 million today when adjusted for currency inflation. After the Extended Fujita scale was developed in 1971, the storm was classified as an overall "EF4", the second highest rating on the scale based upon the specious argument that "…no one could know how well the houses were anchored…". I say specious because the houses built in New England are just as sturdy as houses built in any other part of the country and certainly not as flimsy as the mobile homes seen crushed in many cases. The tenement houses of Worcester did not flop during the (frequent) 60 to 70 MPH gales that accompany Nor' Easters and blizzards, or the North West gust fronts that ripped through several times each year. The houses were built strong to withstand the weather. It is my guess that the Weather Bureau (aka NOAA) were just expressing "sour grapes" about the "…big one that we missed…".

INTRODUCTION: MY TAKE ON WHY THE STORM TOOK ITS PATH

I lived in Leominster, MA in 1953. The "Center to Center" distance from Leominster to Worcester was 20 miles, almost due South. Normally cold front attendant thunderstorms would cross Leominster from the West or Northwest. There were few, if any storms, that came from the Southwest because of one saving grace, the unique alignment/location of mountain peaks in the central part of the state.

An examination of the topography commanding the path of the storm leads me to hypothesize that mountains dictated the direction the storm would eventually take. In the same way that rocks in a stream shunt the water to one or the other side (and produce eddies opposite to the direction of flow). Bring out the topographical maps and note how "rocks" in the airstream were "created" by a series of mountains in the oncoming path of the storm. The first feature listed below created a "slot" to move the storm south as it passed over the hills of the Berkshire range from the Schenectady area.

- Mount Greylock is the highest natural point in Massachusetts at 3,491 feet (1,064 m). Its peak is located in the northwest corner of the state in the western part of the town of Adams (near its border with Williamstown) in Berkshire County. STORM SHUNTED TO THE SOUTH.

After traversing the "high ground" from the Connecticut River to the Quabbin reservoir area other features came into play:

- Mount Wachusett @ 2,006 feet elevation at summit, is located in the towns of Princeton and Westminster in Worcester County, Massachusetts. It is the highest point in Massachusetts east of the Connecticut River. STORM BLOCKED BY THIS PEAK, SHUNTED SOUTH.
- The Crow Hills, located in Massachusetts' Leominster State Forest 2.5 miles northeast of Mount Wachusett, are a single Monadnock with a twin summit, 1,234 feet (376 m) and 1,220 feet (370 m), and a high eastern cliff. FURTHER BARRIER TO ANY "NORTHWARD" SWING.

The supercell was guided south (not only by upper-level winds), but by a virtual "dam" of higher terrain that is angled across the southern Vermont\New Hampshire area.

- Mount Watatic is an 1,832 foot (558 m) monadnock located on the Massachusetts-New Hampshire border, at the southern end of the Wapack Range of mountains. THE FIRST PART OF A "WALL" OF HIGHER TERRAIN TO THE NORTH.
- Pack Monadnock or Pack Monadnock Mountain 2,290 feet (700 m), is the highest peak of the Wapack Range of mountains and the highest point in Hillsborough County, New Hampshire. TERRAIN WALL TO THE NORTH.
- North Pack Monadnock or North Pack Monadnock Mountain is a 2,276-foot (694 m) monadnock in south-central New Hampshire, at the northern end of the Wapack Range of mountains. TERRAIN WALL TO THE NORTH.
- Mount Monadnock, or Grand Monadnock, at 3,165 feet (965 m), Mount Monadnock is nearly 1,000 feet (300 m) higher than any other mountain peak within 30 miles (48 km) and rises 2,000 feet (610 m) above the surrounding landscape. TERRAIN WALL TO THE NORTH.

These mountain features not only defined the choice of travel for the supercell – by creating channels of "least resistance", but air currents ahead of the storm collided with this rising topography and the eddies created may have added additional "spin-energy" to the storm. There is anecdotal evidence to support this theory in the article: ("Study Suggests Topography Can Strengthen Tornadoes", By Ben Benton, October 25, 2013, Insurance Journal)

INTRODUCTION: MY FAMILY CONNECTION

My father was the only boy in his family. At the time of the tornado, four of his sisters were members of the Venerini Order of Catholic nuns and, coincidentally, stationed in Worcester (Edwards Street). It was not long after the boiling black mass of clouds had retreated further to the South and East that the local radio stations began broadcasting that some sort of severe wind storm had created large scale damage in the Worcester area. Out of two concerns, we had witnessed the passage of the supercell formation and there was no phone service into the affected area that evening, my father decided to go into the city as soon as possible and check on his sisters' welfare.

Travel into Worcester by car was limited to emergency personnel and Worcester residents, but (oddly) there was bus service – buses did not "stop" in the damaged area, but proceeded on the main road (at that time Route 12) through to the city center. I was able to go with my father, the next day, on one of the first buses into the city.

I had taken the bus trip before to visit my aunts, that I was certain. This trip was one I would never forget. It plays in my memory like a movie – even today. The day was brilliantly sunny, the sky so blue it hurt. There was no stifling heat and humidity, it was warm, but not uncomfortable. The bus was a "modern" gasoline engine behemoth that had no sound deadening insulation (or springs for that matter). I sat next to the open window and until we reached the city limits everything seemed untouched. There were a few broken branches here and there and a tree was blown down on a lawn, but there was no indication of what was to follow.

We were stopped just at the city line and we were "inspected" by an armed National Guardsman who asked each passenger what "business brought them to Worcester" that day. My father asked about the Venerini convent and was visibly relieved to be told that the storm had completely "missed" that section of the city. We were allowed to proceed, although one person was removed from the bus after protesting that he had a "right to walk through the affected area."

The bus rolled on. It was apparent that the main road (Route 12) had been "Plowed." There were tangled piles of disjointed debris and trees on the sides of the road. Apparently it had been decided to open the road by pushing everything to the sides for the time being. We were approaching the hillside which had been dominated by the imposing structure of Assumption College. The scene changed abruptly.

I remember seeing a chain link fence curled into a crazy spiral and the debris was becoming, not just sporadic piles pushed aside, but a complete wall on either side of the road. I saw the expansive lawn where I knew the brick edifice of the Assumption College should be. My eyes widened when I realized that the Assumption College main building was no more! There, in the brilliant noon-time sun was a crushed, smashed, demolished main building. Each and every window was blasted out, leaving ugly sores across the front of the remaining walls. Everywhere there was "building materials" strewn on the broad lawn…chaos, just chaos. I thought to myself, "Those were brick walls, smashed…" and they were, as the top two floors and the steeple were obliterated… "Those were brick walls." For the moment I was stunned until I saw pile upon pile of tenement houses push down or blasted open with lumber everywhere. The passengers on the bus seemed to gasp and I lost the sound of the bus itself as I was mesmerized by the destruction. Yet, within a mile, the havoc tapered off and a little beyond that there was no damage save for a branch or two out of place.

We did not talk about the storm. My father was not always a person to discuss his feelings. I imagine he was as amazed as I was. The visit to my aunts was uneventful and we took a bus home in the dark.

INTRODUCTION: THIS PRESENTATION

I have been faithful to the original report almost "to a fault." The original report was typewritten and formatted according to the norms of 1953. The free-use copy had been "scanned" into a PDF format, and I repurposed that for MS WORD. In doing so, I was forced to change some of the original formatting, but not the text content, language or presentation style. As an example of minor text changes is that I added "Table" numbers to make searching and indexing easier. I (also) changed the font type, in certain paragraphs, to a "more modern" style for readability. Keep in mind that the original report was typewritten entirely in Courier font!

The photos and maps are reproductions of the originals and I cannot vouch for their quality. If there are "typos" they are intentionally left in as part of the original. Also, the language is the "vernacular" for that time period and it may seem archaic to anyone not familiar with those norms.

My choice for comments, was to use ENDNOTES. This has prevented conflicts with the original footnotes in the document.

The intent is to present the original report "away from" the endless retelling of personal stories and dramatizations that have been so prevalent in attempts to capture the horror and fury of the storm.

As a final note, I must admit that when I first read the original report I found it to be closer to the truth as I knew it, as I remembered it, and the presentation was not "Hollywood."

ABOUT: Anthony F.C. Wallace – The Original Author of the 1953 report

Anthony F.C. Wallace, in full; Anthony Francis Clarke Wallace (born April 15, 1923, Toronto, Ont., Can.), the son of the historian Paul Wallace, and did both undergraduate and graduate work at the University of Pennsylvania, where he was a student of A. Irving Hallowell and Frank Speck.

Wallace was a psychological anthropologist and historian known for his analysis of acculturation under the influence of technological change. Wallace specialized in Native American cultures, especially the Iroquois. His research expresses an interest in the intersection of cultural anthropology and psychology. He is famous for the theory of revitalization movements.

Wallace received his Ph.D. in 1950 from the University of Pennsylvania in Philadelphia and taught there from 1951 to 1988. His students included the anthropologist Raymond D. Fogelson.

His most important work, *Rockdale: The Growth of an American Village in the Early Industrial Revolution* (1978), is a Psycho-anthropological history of the Industrial Revolution. Wallace studied the cultural aspects of the cognitive process, especially when it involves the transfer of information during periods of technological expansion. In other books he compares religion as a movement of "social revitalization" among the American Indians and in modern times. His books include *King of the Delawares: Teedyuscung, 1700–1763* (1949), *Culture and Personality* (1961, rev. ed. 1970), *Religion: An Anthropological View* (1966), *Death and Rebirth of the Seneca* (1970), *The Social Context of Innovation* (1982), *St. Clair: A Nineteenth-Century Coal Town's Experience with a Disaster-Prone Industry* (1987), and *The Long, Bitter Trail: Andrew Jackson and the Indians* (1993).

He was also for a time the Director of Clinical Research at the Eastern Pennsylvania Psychiatric Institute.

He died on October 5th, 2015.

ORIGINAL TITLE: Disaster Study Number 3

Tornado in Worcester, Massachusetts: An Exploratory Study of Individual and Community Behaviors in an Extreme Situation

Original Author:

ANTHONY F. C. WALLACE, PhD.

The Committee on Disaster Studies (1953)

MEMBERS

Carlyle F. Jacobsen, Chairman
A. Dwight W. Chapman, Co-chairman
Charles W. Bray
John A. Clausen
John P. Gillin
J. McV. Hunt
Irving L. Janis
C.M. Lout (Chairman, Sub-committee on Clearinghouse)
John H. Mathewson
Russell W. Newman
John W. Raker
John P. Spiegel

MEMBERS, ex officio

Harry Harlow
Clyde Kluckhohn
Harry L Shapiro

STAFF

Harry B. Williams, Technical Director
Charles E. Fritz
Luisa Fisher
Jeannette F. Rayner
Mark J. Nearman
Helen McMahon

DIVISION OF ANTHROPOLOGY AND PSYCHOLOGY

Harry Harlow, Chairman
Glen Anch, Executive Secretary

COMMITTEE ON DISASTER STUDIES (1953)

The Committee on Disaster Studies is a committee of the Division. of Anthropology and Psychology, National Academy of Sciences-National Research Council. It was established as the result of a request made of the Academy-Research Council by the Surgeons General of the Army, the Navy, and the Air Force, that it: "…conduct a survey and study in the fields of scientific: research and development applicable to problems which might result from disasters caused by enemy action…" [i]

The function of the Committee is to aid in developing a field of scientific research on the human aspects of disaster. The Committee maintains a clearinghouse on disaster research, publishes a roster of scientific personnel in the field of disaster research, and issues periodically a Newsletter. It makes modest grants to encourage research in disaster studies, advises with responsible officials on problems of human behavior in disaster, and from time to time issues reports on the results of disaster research.

At present its activities are supported by a grant from the Ford Foundation, and by special grants from the National Institute of Mental Health of the Department of Health, Education and Welfare, and from the Federal Civil Defense Administration.

Disaster Study Number 3 Committee on Disaster Studies
Division of Anthropology and Psychology

An Exploratory Study of Individual and Community Behavior in an Extreme Situation

TORNADO IN WORCESTER

by

Anthony F. C. Wallace, PhD. University of Pennsylvania

Publication 392
NATIONAL ACADEMY OF SCIENCES – NATIONAL RESEARCH COUNCIL
Washington, D. C.
1956

FOREWORD (ORIGINAL)

This is the third in a series of disaster study reports to be published by the Committee on Disaster Studies. This series is designed to make the findings of disaster research more accessible to research workers and to agencies and officials concerned with disaster problems. It includes studies which have been completed for some time but which have not been previously published, as well as recently completed studies.

The study reported herein was supported by the Committee under Contract Number DA-49-007-MD-Z5 6 between the Department of the Army and the National Academy of Sciences-National Research Council.

A devastating and unprecedented tornado struck Worcester, Massachusetts, and several nearby towns on June 9, 1953. The Committee conducted and sponsored several limited investigations of different aspects of the disaster. Other limited studies were also made. (Dr. Wallace lists these various studies in this report.)

The Committee turned all the data it had been able to collect over to Dr. Wallace and asked him to prepare an integrated analysis of the Worcester tornado disaster. His analysis is reported herein. His development and systematic application of the concepts of time phases and spatial zones in disaster, and his formulation of explanatory hypotheses – the disaster syndrome and the counter-disaster syndrome, the length of the isolation period, and the cornucopia theory of supply – are proving useful both in the extension of theory concerning human behavior in disaster and in sharpening up problems for rewarding research.

The issuance of this report does not necessarily indicate concurrence of every member of the Committee on Disaster Studies in every statement made in the report, nor does publication imply Department of Defense indorsement of factual accuracy or opinion.

Carlyle F. Jacobsen Chairman
Committee on Disaster Studies

PREFACE (ORIGINAL)

The Committee on Disaster Studies of the National Research Council, during the spring and summer of 1953, undertook to carry on a series of exploratory field studies of disaster. Some of these studies were conducted by members of the Committee staff, some by persons retained as consultants, some by organizations under contract.

These studies involved many aspects of disaster: evacuation experiences, communication, rumor, panic, rescue and rehabilitation, etc.

Following the tornado which struck Worcester, Massachusetts, on June 9, 1953, several organizations sent personnel to study various phases of the disaster, particularly what happened during the rescue and rehabilitation period. The Committee itself sponsored or facilitated: a study of communications in Worcester and in Flint, Michigan (struck by another tornado the day before) by Irving Rosow, a graduate student at the Russian Research Center at Harvard; a study of the role of the Catholic Church, in Worcester and Flint, by a team from the Catholic University of America, including Carroll Brodsky, John Muldoon, and Regina Flannery Herzfeld; two studies of medical care following the disaster, one by the Massachusetts General Hospital, and the other by Jeannette Rayner of the Committee's staff; a study of the psychological and physiological effects of the stress of their tornado experience on a group of previously studied Worcester firemen and industrial employees; and a brief "horseback survey" by the writer aimed at identifying spatial and emotional dimensions of the disaster which would repay systematic analysis later in this or an analogous situation. Other studies made by a variety of persons and organizations are listed in the bibliography of this report.

These studies, and other sources of data, have been collated in the preparation of this report, which is an attempt to analyze the Worcester tornado disaster as an event, according to the categories of the theoretical model developed in the introduction.

It is avowedly exploratory, as the sub-title indicates, in two senses: the conceptual formulations (the time-space model, the disaster syndrome, the counter-disaster syndrome, the isolation period, and the cornucopia theory) are intended to stimulate thought rather than to make converts to a system, and accordingly are not presented as a formal body of theory nor as a set of hypotheses verified adequately by the available Worcester materials; and the empirical data, almost entirely compiled by other observers with a variety of interests, are evidently uneven in quality, quantity, in representativeness, and in precision of reference to the matters I have chosen to emphasize. Furthermore, such generalizations as I have made or implied about disasters as types of event obviously will require the assembling of much comparative materials, both for general validation and for necessary qualification.

With these caveats in mind, it is hoped that both the empirical data on the Worcester case, and the theoretical formulations which have been worked out to organize these data, will be of some use to scholars and administrators interested in disaster studies.

The interpretations and opinions expressed ate my own and do not necessarily reflect the views of the Committee on Disaster Studies.

I should like to make special acknowledgment to several persons and organizations who have helped me greatly in preparing the Worcester report: Harry B. Williams, Technical Director of the Committee on Disaster Studies; W. N. Fenton, until recently the Secretary of the Division of Anthropology and Psychology of the National Research Council; Jeannette Rayner, staff associate of the Committee on Disaster Studies, with whom I worked over at length the disaster syndrome formulation; R. L. Polk & Co., of Boston, who have kindly permitted reproductions of the map of Worcester from their 1953 city directory; and the Worcester Telegram and Gazette and radio station WTAG, who made available tape recordings of their tornado broadcasts, and plates of all photographs used in this report.

Anthony F. C. Wallace, PhD. University of Pennsylvania

INTRODUCTION: *A Time-Space Model of Disaster as a Type of Behavioral Event*

In my memorandum on the Worcester survey, and further in a memorandum on the literature of disasters, I suggested the desirability of approaching the study of disasters with the expectation that a general model of disaster as a type of event could be formulated, with characteristic dimensions in space and time along which various phenomena might be plotted, and that disasters could by systematically compared and analyzed with respect to variation along these dimensions.

At the same time, I suggested the possibility of using the relatively well-described Worcester event as a proper place to apply the early formulations of the model, and, by recurrent attention to the "fit" (or lack of it) between theory and fact, to produce both an improved theoretical model and a. more coherent picture of the Worcester tornado disaster as a. total event.

In earlier publications, Powell, Rayner and Finesinger had developed, from their empirical studies of several disasters of different types, a model of the time dimension in disasters. Disaster time was in this model divided into seven stages; given disasters varied of course in the length, extent of overlap, repetition, and behavioral content of each stage but these variations could be described in terms of the stages themselves. The stages were distinguished by the predominant disaster-related behavior characteristic of persons involved in the disaster; the length in absolute time was variable.

Table 1: Time Stages in Disaster	
Stage	**Function**
0. Pre-disaster conditions	Determining, to some degree, the effect of and response to, impact
1. Warning	Precautionary activity
2. Threat	Survival action
3. Impact	"Holding on"
4. Inventory	Diagnosis of situation and decision on action
5. Rescue	Spontaneous, local, unorganized extrication and first aid; some preventive measures
6. Remedy	Organized and professional relief, medical care, preventive and Security Measures
7. Recovery	Individual rehabilitation and readjustment; community restoration of property and organization; preventive measures against recurrence.

/After Powell, Rayner, and Finesinger, 1953. /

MAP 1:   The City of Worcester

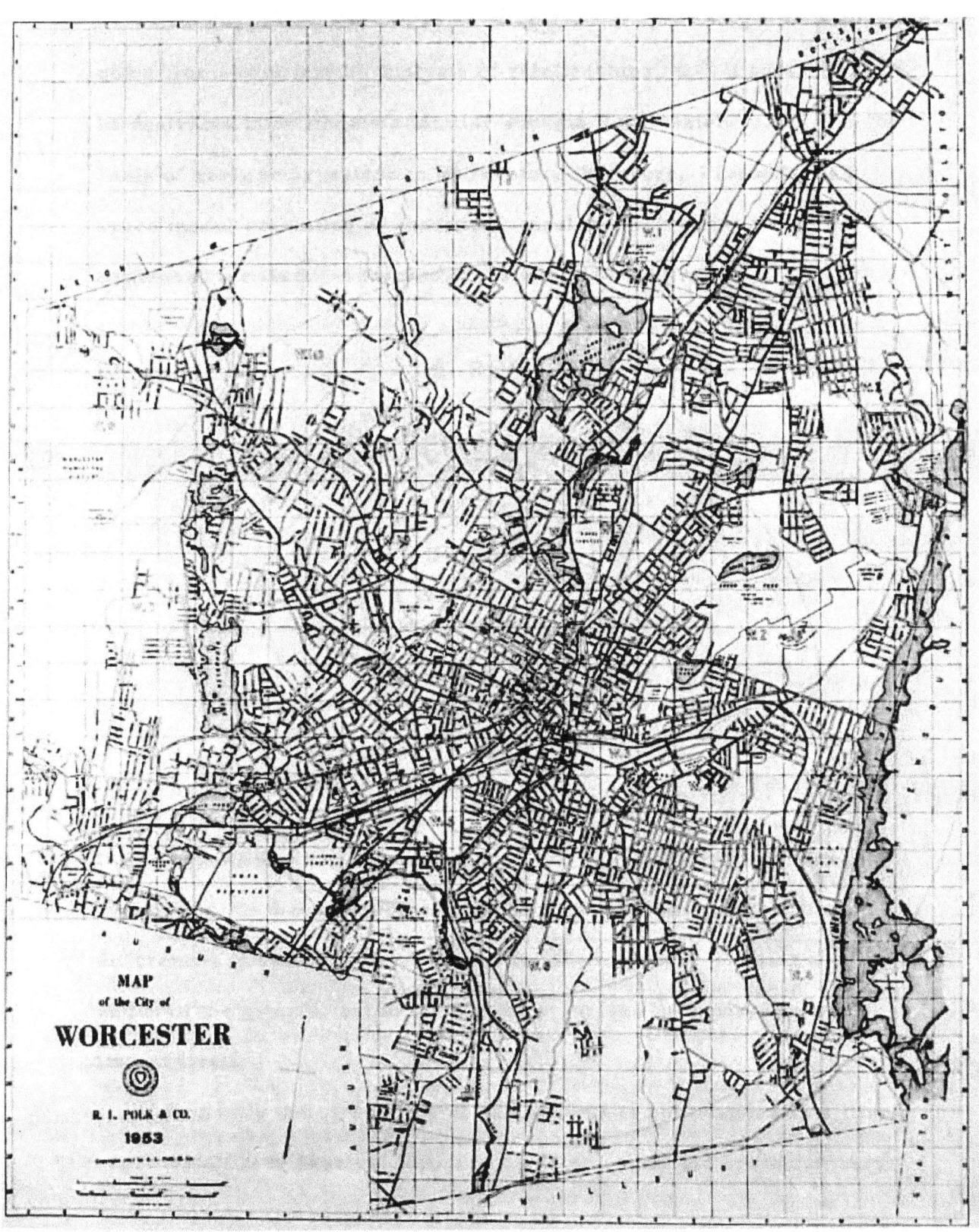

This approach seemed to be so fruitful, in providing a vocabulary and a framework for the analysis of relatioships, that it appeared to be desirable to formulate a similar schema for disaster-space. On the basis of early impressions in Worcester, therefore, I formulated a space model consisting of concentric circles, each circle being distinguished for its most marked disaster-related function:

It was hypothesized that the population and resources of each area play systematically different roles in relation to disaster, and that these differences in behavior are related not only to the pre-existing organization of the ground, but to the spatial organization imposed by the impact itself.

Evidently the circularity of this schema is not intended as a literal representation of physical layout in disaster, since the areas can vary greatly in shape from one disaster to another, and in any one disaster, the areas can vary greatly in width from place to place. The impact area - the area within which physical destruction occurs - is divided into an area of "Total Impact" and an area of "Fringe Impact."

"**Total Impact**" does not mean total destruction but rather the working of the impact agent with the full severity characteristic of it in this disaster.

"**Fringe Impact**" is minimal in comparison with the "Total Impact" and usually involves minor damage and few or no serious injuries. Criteria for drawing the imaginary boundary between these two sub-areas will vary with the type and destructiveness of the impact agent, however: much of the fringe area of a hydrogen explosion might show damage more serious than the total impact area of, let us say, a flood; and often the two areas will shade off into one another on a rather smooth gradient, so that the drawing of a line becomes rather arbitrary.

Even in such cases, however, I suspect that the concept of "Total" and "Fringe" is utilized by victims and rescue workers, even though the drawing of a line would be difficult. In the case of Worcester, because of the nature of the tornado as an impact agent – the fringe area was sharply and obviously demarcated from the area of total impact; one side of a street, for instance, might show only stripped shingles and leafy debris; on the other side, whole houses would be reduced to rubble.

"Filter Area" is, the label attached to that region, broader or narrower, immediately adjacent to the impact area, through which both traffic and information must pass to and from the filter area. Because it is in a sense the narrow end of the communication funnel and because there are likely to be traffic and communication jams in it at points of entry to (or exit from) the impact area, it tends to act as a filter, screening out certain kinds of traffic and information. The filter area is also likely to supply certain services, in the way of communications, rescue, first aid, and comfort, to victims in the impact area before organized community and regional aid arrive.

The areas of **"Organized Community and Regional Aid"** are only selectively affected by the impact, and their services in large part are carried out by various counter-measure agencies, such as police and fire departments, hospitals, relief agencies" etc. A clear line of demarcation between. community and regional is easy to find in the case of urban areas, because the city itself can be defined as the community; but the boundaries of the region are vague, ramifying out through county, state, and major geographical area. Where the national government actively intervenes with more than pre-established administrative facilitation of the work of the regional or community offices of national agencies (for instance, by assembling military personnel from widely scattered points rather than by authorizing through standard operating instructions the participation of local forces), one might speak of an "organized national aid" area coextensive with the national territory. Distinctions between "regional" and "national" are apt to become distinctions between levels of administrative authority rather than valid geographical distinctions, however, because any major disaster, at least in an industrialized society, is likely to involve communications, transport, and industry in all parts of the country.

In the present study, the attempt has been made to bring these two schemas together and to forge a systematic analytical tool by which data may be classified in time and space. No major modifications have been made in the original time-schema of Powell, Rayner, and Finesinger. In the chapter-by-chapter presentation of the Worcester material I found it to be more convenient to treat the warning and threat periods as indistinguishable, because of the slightness of the warning itself. The post-impact periods of Powell, Rayner, and Finesinger (inventory, rescue, remedy, and recovery) have been somewhat consolidated and renamed, chiefly to relate the time zones to space zones.

Thus my "isolation" period is equivalent to Powell's "inventory" and "rescue" periods; my "rescue" period is his "remedy" period; and my "rehabilitation" period is his "recovery" period. I have added a final period, "irreversible change," to take account of the fact that recovery or rehabilitation is never complete, and that the original equilibrium state before the disaster is not fully restored; a slightly different equilibrium is attained. The essential elements of the conceptual model of that type of disaster in which a lethal impact is relatively sudden, brief, and sharply defined in area, is as follows:

1. STEADY STATE

This is the system of regular energy-distribution (action) obtaining in all of the ultimately affected areas at the moment just preceding the warning period. The system will probably be in equilibrium, or nearly so, at the time of any given disaster. By equilibrium I mean that energy discharges are of a repetitive and predictable nature, in response to chronic stresses; furthermore, such stresses are eliciting effective conventional responses. In other words, the cultural system, and the personalities of the population, are operating sufficiently smoothly to obtain stress reductions for the population, such that the total quantity of stress in the area at large is not systematically increasing or decreasing (although there will be random variation).

Elements in this system of energy-distribution are: terrain, topography, climate; the culture of the population involved (including their security agencies designed to protect them from disaster); certain non-cultural characteristics of this population, including the distribution of various demographic factors, and the distribution of personality types ("national character"). It should be noted, however, that many of the characteristics of the system are cyclical, so that at any given moment the actual distribution of elements may be different from that at another (for example, the alternation of night and day, the summer migration from the city to vacation spots in the hinterlands). Furthermore, there is a sort of "random incidence" of stresses and responses thereto, such as births, deaths, business openings, the making and breaking of friendships, the development of gradual trends in the system itself, and so on, which may mean that the momentary situation may be very different at successive times within the same steady state.

Thus both the total system and the momentary situation at the moment of warning, threat, or impact are important determinants of what happens as the disaster proceeds. To be a disaster, the impact must upset the total system; but the degree and nature of upset which a given impact produces depends on the momentary situation of that system.

2. WARNING

This is the presentation of generalized cues to the energy-distribution system. This may occur over a long period of time. These cues may be meaningful or not, depending on previous learning, the state of perceptive-cognitive apparatus, and the nature of the communication network within the energy-distribution system. The cues themselves may be susceptible of several interpretations; they tend to be ambiguous. The cues may be widely presented to many units of the system or to a few. There may or may not be protective action taken as a result of the cues.

Warning, as distinguished from threat, does not specify certainty of impact at any particular place at any particular time, but rather a probability greater than normal within the system that such an impact will affect a particular place within a range of time: in other words, the cue - if correctly read - means, "such and such" a general kind of impact, severity not fully predictable, may hit within a given span of "time."

As a corollary of the above, during the warning period there is no differentiation of disaster space into the five functional areas.

3. THREAT

The threat period also involves the presentation of cues to the disaster region. Threat, however, involves cues which, to many persons at least, are not ambiguous. The period of threat is usually short, being followed either by impact or by awareness of the danger having passed. Threat involves the presentation of a cue which is unmistakable, and requires the operational assumption of certainty of impact, even though hope of escape may persist.

An initial differentiation of the disaster region begins, into a "threat area" (areas 1, 2, and 3 in my diagram) and a "continuation of warning" area (4 and parts of 5). Many persons in the threat area know what it is, know approximately when it will be, and assume that it will hit them. Consequently, some emergency survival action (flight, retreat to cellars, etc.) is usually taken in the threat period.

The threat cues, however, like warning cues, may not reach all persons, and may not be interpreted as threats if they do reach a given point. What is a threat of immediate impact to one person, may be a casually-received warning to a second, and may not be recognized as either a threat or a warning by one third?

4. IMPACT

The impact period is the time during which physical destruction is being accomplished by the impact agent. The destruction accomplished during the operation of this agent (whether it be tornado, flood water, hydrogen explosion, or whatever) is called the primary impact.

The primary impact is only that damage inflicted during the action of the agent; the consequences of this primary impact are occasionally irreversible, unchangeable and final, but usually primary impact is followed by a secondary impact which is susceptible to manipulation. Secondary impact properly pertains to the following period, but for clarity of exposition will be discussed here. Let us take an example. The driving of a piece of glass into a man's arm, severing an artery, during the thirty seconds of the tornado's passage over an area, is primary impact. At the close of the brief impact period (i.e., when the tornado is past) the man is left with a cut arm, blood spurting out of it. The bleeding which follows, and which may, if unchecked, cost the man his life, is secondary impact. The distinction is arbitrary, in this instance, from the standpoint of the traumatic process, but is made in order to clarify the time analysis. What happens of a destructive nature during the operation of the disaster-agent is primary impact; destruction which follows in time is secondary impact.

The impact period differentiates the "threat area," mentioned in the preceding section, into three other areas. The impact area itself is divided into two parts, the areas of "total impact" (where death and serious injury and property damage occur) and of "fringe impact" (where only minor injury and property damage occur). Theoretically, depending on the characteristics of the disaster agent, several sub-areas could be defined, distinguished by nature and severity of impact. The filter area, just outside the impact area, is also partially defined during impact: some of its population become aware that they are "near-misses."

5. ISOLATION

During the isolation period, the impact area is "isolated" in the sense that personnel and equipment from areas 3, 4, and 5 have not come in to perform rescue, first aid, reassurance, and evacuation functions. Strictly speaking, the length of the isolation period will vary according to proximity to the filter area and to roads, and according to other chance factors, so that some points will not receive aid for perhaps half an hour, while some will be swarming with helpful neighbors in the filter area, ambulances, and fire apparatus within a minute. "The isolation period" therefore, is an abstraction from the sum total of all the isolation times of all points in the impact area. When one says "the isolation period" lasted for such and such a length of time, one is expressing one's estimate as to some measure of central tendency of the distribution of all isolation periods within the impact area or a part of it.

The isolation period is the time during which secondary impact achieves its chief ravages. Secondary impact includes all those destructive and deteriorative processes which have been unleashed by the primary impact: in regard to trauma to individuals, bleeding, infections, interruption of normal breathing, physiological shock, brain injuries, etc., etc.; it also includes such phenomena as exposure to inclement weather, various emotional stresses, and possible (if the isolation period is prolonged) deprivation of food, water, and sleep. Secondary impact also includes such things as the hazards of live wires; fires and conflagrations; leakage of poisonous or explosive gases; spread of epidemic disease as a result of inadequate sanitation, food, enforced crowding, etc. Many of the more critical and dangerous secondary impacts, however, are being reduced by the impact-area population during the isolation period, and the arrival of organized aid from the community and region sharply reduces secondary impact. A secondary impact may, of course, if it gets out of control (as in a firestorm) become a new primary impact agent, or can be regarded as a continuation of primary impact.

As is indicated in Powell, Rayner, and Finesinger's functional analysis of disaster time, the impact area population has two tasks during the isolation period: inventory and "rescue" (which term covers not only rescue per se, but a variety of counter-measures designed to terminate or reduce secondary impact). The length of time taken in inventory is probably highly variable from person to person, and as will be indicated later on, the taking of inventory (with the attainment of such perceptual and cognitive reorientation that a decision on action is possible), is not always completed by the time aid arrives from the outside, so that some individuals are found in a more or less "dazed" condition, while others are relatively well-oriented and effective in action.

While the impact areas are "going it alone" during the isolation period, the community and the region are in process of learning that a disaster has happened, of making inventory and reconnaissance of the extent and nature of primary (and to some extent secondary) impact, and of mobilizing and dispatching the security forces within reach of the available communications network. The filter area further differentiates from the regional and community aid areas, being the only area which has reasonably adequate awareness of what has happened, partly as a result of having seen and heard what happened, and partly because into it the first evacuees are passing, and in it traffic jams are developing as vehicles pile up at the edge of the impact area. Notification and mobilization in the community and region are selective during this period: security agencies and personnel connected with the communications system know what has happened and are mobilizing, but most other persons and organizations know very little.

6. RESCUE

The rescue period, at any point, begins when aid arrives from areas 3, 4, and 5. During this rescue period, a combination of three groups of personnel – impact area survivors, unorganized spontaneous volunteers chiefly from the filter area, and organized security units – work to combat secondary impact within the impact area. (In some parts of the impact area, filter area volunteers arrive before the organized security units; in other parts, organized security units are the first to arrive.) Particularly important are the roles of the police and fire departments, or other security forces (such as military units) who are on a 24-hour operational basis. The chief functions to be performed are extrication, first aid, evacuation, emergency hospital medical care, and termination of such secondary impacts as fire and "hot" wires.

During the rescue period the five functional areas reach maturity in the differentiation of their separate roles with respect to the disaster. There is a decreasing frequency and degree of personal emotional involvement with the disaster in the populations of areas 3, 4, and 5, respectively, and an increasing tendency for involvement to be highly selective and aimed at organizations rather than populations.

7. REHABILITATION

When secondary impact has been reduced to a minimal point – in other words, when equilibrium has been temporarily reestablished, even if on a level the society defines as undesirable – effort will be made to bring the system back to the original state. This involves extensive welfare and reconstruction activities; this is the longest period, the most expensive period, and the one which involves the greatest number of organized aid-personnel. The organizations responsible for the rescue operations tend (with the exception of hospitals and medical personnel generally) to reduce their involvement, and new organizations – the Red Cross, government relief agencies, and insurance companies – to take over responsibility.

The rehabilitation period in a sense may be said to last indefinitely, but for simplicity in analysis it is preferable to set a time limit – let us say, arbitrarily, one year - and to consider whatever is not yet rehabilitated by that time to have become, in its changed condition, a part of the new "steady state" or equilibrium system.

The lines separating the various areas of disaster space begin to blur during this period, and toward the end of it virtually disappear as whatever residual changes produced by the disaster become a part of a new system in which functional relationships (e. g., financial, medical, repair, etc.) depend less and less on location with respect to the impact area.

8. IRREVERSIBLE CHANGE

As has been indicated, eventually (let us say, a year later) a new steady state will have been established somewhere between the situation at the end of secondary impact and the pre-disaster equilibrium system, toward the reestablishment of which rehabilitation functions were aimed. The five areas are no longer functional; they exist as memories.

It should be made clear that "irreversible change" means a change in the system, not merely a change in the units of which the system is composed. People may die, lose their money, etc., and not affect the system, if others replace them; but if a significant change in the population pyramid, or in the occupational structure, or in the standard of living, or in the organization of an agency like the Fire Department is left, the system has changed irreversibly.

In addition to presenting the story of the Worcester tornado along the lines of the analysis outlined above, several particular topics will be discussed: the **"*disaster syndrome*"** (a name for a complex of evolving attitudes and overt behavior, one of whose manifestations is the frequently referred to "apathy" of survivors from areas 1 and 2); the **"*counter-disaster syndrome*"**, displayed by persons in areas 3 and 4 (rarely from 5) who feel responsible for mitigating the effects of the disaster (characterized by over-conscientiousness and competitiveness in the rescue and early rehabilitation stages); the importance of the length of the isolation period; and the **"*cornucopia theory*"** of disaster relief and civil defense.

After going through a considerable amount of material on the Worcester tornado, I am left with rather a humble feeling. I have been forced to consider and evaluate as best I can, in terms of my own theoretical approach, the actions and standard operating procedures of professional disaster workers and organizations; and I am acutely aware that I have probably over-simplified and at times possibly misinterpreted situations and procedures. I have attempted to keep constantly in mind that research into human behavior in extreme situations is in an exploratory state and that it is important to develop a language, a body of concepts, and a theoretical structure to which problems can be defined, and data selected and analyzed. But as must be true of introductory research in any new field, the researcher is apt to be somewhat naive in his approach to some of the empirical materials. The cardinal example of naivety in this study is the failure by myself – and to a greater or lesser degree by other investigators – to recognize the primary role played by the police and fire departments. Although a few firemen were interviewed by Powell, they were selected because they had participated in an earlier study of physiological responses to frustration, and were not interviewed primarily as informants on the events of the isolation and rescue periods. Not one interview with a policeman have I seen, although I believe Rosow interviewed some police officials concerning communications. Administrative officials of the police and fire departments similarly were virtually ignored. Even the press tended to overlook the regular city security agencies (police, fire, and public works) in favor of Civil Defense, the Red Cross, and the National Guard (who were much later on the scene).

If I were doing this study over again, I would concentrate on police, fire, and public works department personnel first and most intensively for the isolation and rescue periods, and I would recommend that in other, similar studies, these personnel be very quickly and thoroughly interviewed. I would also tend to be extremely cautious in accepting criticism of these and other standard agencies, because of the tendency for the excited and conscientious volunteer auxiliary to discover faults in the procedures of the more phlegmatic professionals.

In future empirical studies, it would be desirable to interview a combination of expert informants and representative informants chosen according to an appropriate sampling design and by the use of an interview guide defined in advance for relevance to matters of theoretical importance. Many of the basic phenomena of disaster behaviors have been isolated and defined, and a conceptual structure exists for their classification and analysis. It would be desirable now to learn the distribution and dynamics of these phenomena – such as the disaster and counter-disaster syndromes – by systematic and meticulous study.

I. STEADY STATE: The City Before Impact

1. General Characteristics

Worcester is the second largest city in Massachusetts, being exceeded in size and importance in the state only by Boston. It lies on the west bank of Lake Quinsigamond about forty miles west of Boston, with which it is connected by a super-highway. Worcester calls itself "The Heart of the Commonwealth." It is a community which, in the course of the past two hundred and fifty years, has slowly developed a rather high level of civic responsibility and morale, of flexibility in meeting new situations, and of identification with a cosmopolitan tradition while at the same time maintaining something of a "small-town" atmosphere. This high level of morale is, I believe, most important to keep in mind in evaluating what happened on June 9, 1953. [ii]

It is, of course, difficult to define and measure objectively such phenomena as the "morale," "cosmopolitanism," etc., of a city. Nevertheless, it has been my distinct impression, both in the course of two brief visits to Worcester (after the tornado) and from my reading, that Worcester is a city of relatively high urban morale, and that this may well have had a great deal to do with the surprising effectiveness of spontaneous, unplanned, and uncoordinated rescue and relief operations, both by individuals and by official agencies. The ratio of deaths to number of dwellings destroyed was noticeably lower in Worcester than in Flint, Michigan, and a difference in local morale may have had something to do with it. [iii]

The city lies in the hilly, lake-studded country of central Massachusetts. Within the city itself there are over a dozen small lakes, reservoirs, and "ponds," some of them associated with parks; the many steep hills scattered through the town rise to heights of up to 400 feet above the lowest parts of the town. The hills and the little lakes, and the antiquity and topsy-growth of the town itself, have resulted in a street layout that is extremely irregular and arbitrary; while individual subdivisions and development areas have a gridiron or other through-street plan, and there are several arterial streets and high-ways, there is a very high frequency of dead end streets, T and Y forks, and unexpected changes in street names which made movement difficult for me as a stranger and must inevitably tend to constrict traffic flow even for local residents. A well-informed local resident, in fact, told me that the "main civic problem" was "traffic," the difficulties arising from the rarity of through streets, and that a hill on the edge of Main Street threw too great a burden even on that major artery[1]. The strangulation of through traffic may be in part responsible for the tendency for subdivisions and developments to retain a local name and identity: being functionally relatively isolated, they do not become so quickly merged into a larger gridiron- whole. Thus, areas of a few squares will retain a definite name in the city's terminology: e.g., "New Worcester," "Indian Hill," "Fairmount," "Greendale," "Union Hill." One might speculate that in consequence there may be somewhat less of the phenomenon of urban anonymity in Worcester than in some other cities; people are not merely residents of "Worcester," or of "South Worcester," but of little

---

[1] Kranich, conversation, 19 June 1953.

communities each with its own name and identity.

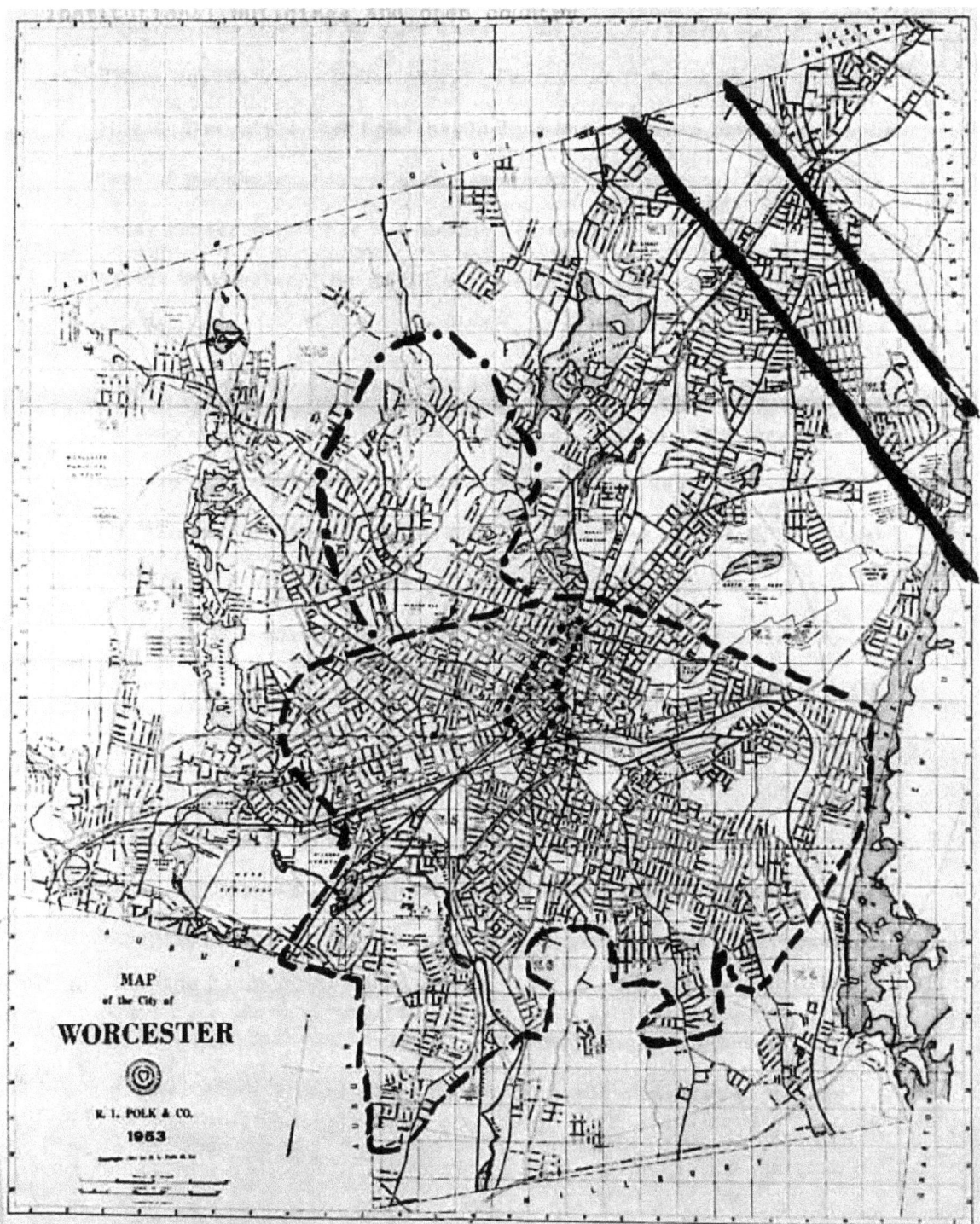

MAP #2: Ecological Zones in Worcester

── Path of the Tornado
⋯⋯⋯ City Center
─ ─ ─ "Working Men's" homes with some industry and public buildings
─ · ─ · ─ "Upper Class Housing"

Unmarked area in Map #2 is a mixture of new middle-income housing, industry, institutional buildings and open country.

As can be seen from the map, in spite of the peculiarities of the street pattern, the city can be divided into roughly concentric zones: [iv]

(1) "Downtown," including the City Common, the City Hall, the Post Office with federal agency offices, newspaper and radio offices and studios, some of the "better" shops and department stores, amusements, churches, Municipal Auditorium (where Civil Defense Headquarters were located), and various business offices and headquarters;

(2) A small slum area on the edges of "downtown;"

(3) A ring of working-peoples' homes, among which are scattered a variety of public and private institutions: Union Station, hospitals, colleges, athletic fields, etc.;

(4) The "best residential district," featuring expensive homes, large lots, privacy, lack of through traffic, and closeness to the city center;

(5) Industrial sites, located on the major transport arteries;

(6) Relatively new, middle class housing developments and public housing projects, located in fairly open country.

Surrounding Worcester city are a number of more or less suburban towns in Worcester County which regard Worcester as a commercial and administrative center; Worcester thus "serves" a population of perhaps 600,000.

The population of Worcester in 1950 was 203,486; in 1940 it was 193,694.[v] Eighteen per cent of this population was foreign born. The national origins of the foreign and native born, taken together, are: English (the original source, of course, in 1713 when the first permanent settlement was made); Irish; Swedish (with a few Norwegian and Finnish); and French Canadian. There is also a small group of Armenian, Albanian, and Syrian origin, to some extent localized, according to report, in and near the slum area south of "downtown," and there is said to be a "Jewish section" in the northeast, part of which lies in the "best" residential district. There are few Negroes [see Endnote iv.] and a few Polish and Italian immigrants. It was impossible to obtain information on problems of inter-group relations in the city, but I heard one Protestant informant express fears that the Catholic church organization might attempt to use the disaster as an opportunity for advertising itself (since the priests wear a "uniform," it was feared that their service would be more visible than those of the Protestant clergy). Possibly the fact that many of the Catholic church organizations in Worcester minister to French-speaking immigrants from French Canada and consequently use French in at least one church and at Assumption College, has tended to maintain their separateness. But I saw no great evidence that ethnic or religious differences were reflected in much hostile social discrimination in Worcester. The city council, for instance, consists of Messrs. Holmstrom. Sweeney. Duffy, Katz, Marshall, O'Brien, Rousseau, Soulliere, and Wells[2].

Worcester's industrial development began in the first decade of the 19th century, when local

---

[2] Field notes, AFCW, 15-19 June 1953; Polk, 1953, 9.

manufacturers (with a local population of 3900) produced clocks, textiles and textile products, paper and wood products, machinery for the textile works, etc. Since that time Worcester has continued to have a diversified industrial base, which at the present time includes some 700 manufacturing establishments employing over 50,000 persons in such industries as steel and wire (e.g., American Steel and Wire, a division of United States Steel, in South Worcester); abrasive products (e.g., Norton Company, in north Worcester); iron, steel, and special metals products; machine tools (e.g., Reed-Prentice, south-western Worcester); leather (e.g., Graton and Knight, eastern Worcester); and so on.

The bulk of the manufacturing and processing is in the southern half of the city but a finger of industry runs up West Boylston St., one of the arterial roads, in northern Worcester. Commercial establishments employ about 30,000. The city was hit hard by the collapse of the textile and shoe industry in New England after World War II. Worcester machine tool companies supplying the shoe-and-textile trade had to cut down, and about 1949 there was widespread unemployment.[vi]

A forward-looking Chamber of Commerce, however, has been making attempts to diversify industry even more and to reduce a tendency to dependence on machine tools. Probably also in response to the economic troubles of Worcester, and under the leadership of the city's businessmen, a reform administration came into power about 1949. It would seem that the city's industrial leadership has been planning and carrying through a vigorous program of action in response to the severe threat to the economy of New England's cities. The business of the building trades in Worcester (there has been a good deal of residential construction) and the fact that there has even been some industrial expansion recently, are evidences of the morale I spoke of before.

Thus, for instance, the Norton Company (owned by local families) had just completed an addition to their plant in June, 1953; the electric company had also recently expanded; the Housing Authority had just put up a new housing project; and extensive redevelopment and traffic reorganization of the downtown area is planned. It would seem, in other words, that Worcester had, by June, 1953, begun to emerge from the threat of economic disaster. To some extent also this had been accomplished without disruption of the pattern of ownership of industry by local families, even though a subsidiary of U. S. Steel (American Steel and Wire) had moved in.[3]

Although intra-city traffic is poorly organized, as I have indicated, transportation to other places is well provided. There are three railroads; a super-highway links Worcester to Boston; U. S. 20 passes a few miles south of the town, and connects at Hartford with the Wilbur Cross and Merritt parkways to New York City; three airlines use Worcester Municipal Airport, including Northeast and TWA; and the city is the center of a web of smaller roads and highways with attendant opportunities for bus and truck service.[vii]

---

[3] Field notes, 15-19 June, 1953; Polk, 1953.

The city is something of an intellectual center. It is the seat of Clark University, which has the distinction of being one of the pioneer graduate schools in the country, having among other things given Franz Boas, the virtual founder of American anthropology, his first academic position, and having invited Sigmund Freud to deliver here his first American lectures. There are also a Polytechnic Institute and a State Teachers' College, and three Catholic colleges (Holy Cross, Assumption, and Anna Maria). The American Antiquarian Society is located in Worcester, and contains over half a million titles relating to American history. There is a public library containing 485,291 volumes. There is an Art Museum with its own school administered in collaboration with Clark University. [viii]

Politically, Worcester is both a municipality in itself and the seat of Worcester County. The city government is now organized under a "Plan E Charter," and accordingly there is a City Manager, a Mayor with nominal salary (in 1953, this was Andrew B. Holmstrom, vice-president and general manager of the abrasives division of the Norton Company), and a City Council. The town is approximately half Democrat and half Republican in political party affiliation;[ix] the charter plan is intended to provide "non-party" government by insuring proportional representation. I am not informed on the ins-and-outs of the political situation in the town, but had the definite impression that some of the noisy wrangling in city hall which was much publicized after the tornado was an outgrowth of unresolved political conflicts, probably dating from the realignment of power four years earlier.

2. Counter-Measure Agencies

Agencies and organizations, part or all of whose mission is to prevent, mitigate, or relieve physical disaster are to be found in any city. In Worcester itself on June 9th they included: the police department; the fire department; Civil Defense; the local Red Cross chapter; the hospitals and medical personnel generally; City Welfare department; various charitable agencies, including the Salvation Army and various church organizations. Available to Worcester according to existing arrangement or custom, for use in case of need, were State Police; regional and State Civil Defense; the National Red Cross; National Guard units; and a vast and intricate network of private insurance on dwellings and business properties, and of State and Federal funds and facilities under a multitude of bureaus, which could, depending on circumstances, powerfully mitigate the economic effects of disaster. The public utilities organizations (electric power and light, gas, and telephone), while not essentially counter-measure agencies, were disaster-conscious and prepared for counter-disaster action. The state of readiness for a disaster with a single, brief, and unexpected impact of the explicitly counter-measure organizations and facilities was extremely varied. The police and fire departments were (as in all large municipalities) operating on a twenty-four-hour basis, and were manned not by volunteer or part-time help but by professionals, trained for the job and paid salaries by the city.

The police department had under its control almost all of the ambulances in the city; the major hospitals did not control the dispatching of ambulances. Personnel of both fire and police departments were trained to act as disciplined teams; were accustomed to dealing with the consequences of physical violence; and thus were equipped to perform such auxiliary tasks as rescue, first aid, and leadership of unorganized groups and confused individuals. The headquarters of both fire and police departments are in the downtown area close to city hall. The police tend to be more centrally directed than the firemen: the bulk of the control flows from the downtown office, and there are but two outlying stations, for the motor patrol (south Worcester) and for the lake shore district (west).

The nineteen or more fire stations, however, tend to be concentrated in and around the downtown area too, even though there are several some distance out.[x] Furthermore, the police department with its nearly 400 patrolmen and officers, and a number of trained auxiliary police and constables, had a supply of cruiser cars with two-way radio, which made possible considerable mobility and flexibility of communication.[xi]

The Civil Defense organization in Worcester was in an embryonic state, in comparison with the maturity of the police and fire departments as institutions. Civil Defense was conceived - on paper - to be the disaster counter-measure agency par excellence. The plan called for Civil Defense to assume the role of central authority: to mobilize resources, direct field operations, and coordinate existing operating agencies. Not all of the potential disaster agencies were formally and officially committed to accepting CD control[4]. Civil Defense in Massachusetts is organized in three levels: the state level, with headquarters at Boston; the regional level (Worcester was in Region #3); and the municipal level (Worcester thus had its own CD organization). Region #3 and Worcester shared a common HQ office in the Worcester auditorium, with one telephone apiece. The City Manager of Worcester was also the CD director of Region #3.[xii]

On the day when I was shown through the auditorium (June 16, 1953) the civil defense officer in charge, with his telephone, city map, desk, and files, was ensconced directly beneath a large window (which to this naive visitor looked like a sure bet, in event of air attack, to cut to ribbons director, telephone line, maps, and any other staff personnel who happened to be near).[xiii]

The mission of the State Civil Defense (on which Worcester of Region #3 CD could, of course, call) in addition to organizational communication, and educational and training functions, included supplying certain types of auxiliary heavy equipment and specialized service (ambulances, trucks, rescue teams)[5]. The city Civil Defense had received two sets of rescue tools under the Federal Contributions Program, but no method had been devised for their transport and no rescue workers trained in their use[6]. There was no overall system of wardens, but information on "Duck and Cover" (ibid. Endnote ix.) methods of personal protection under air attack had been widely disseminated, especially in the school system, and there had been a civil defense drill

---

[4] Rosow, M. S. 1954 Ch 1.
[5] Rayner, NOTES: Crowley and Gauthier
[6] Knight, MS.

in June, 1953. But as of 9 June, 1953, Worcester Civil Defense, in spite of the energetic efforts of the director had developed only an embryonic organization with physically vulnerable headquarters, and very few trained personnel.

The Worcester chapter of the American Red Cross, with headquarters on Harvard Street, about a block from CD Headquarters, was, of course, the local office of the national organization designated by law as a quasi-governmental agency with responsibility (defined by congressional charter) for aiding government at various levels in the event of disaster. Among the recognized functions of the Red Cross following a disaster is an obligation to assist Civil Defense, under Civil Defense authority, primarily in the provision of emergency supplies of food, clothing, and shelter; if other agencies cannot handle rescue, medical care, information and location services, etc., Red Cross may assist here too. The Red Cross, both locally and nationally, has the peculiar position of having a broad but vaguely assigned public responsibility, with formal cooperation agreements with military, CD, and other disaster relief agencies, but without access to public funds; thus relief, accounting and fund-raising activities must proceed hand in hand. Furthermore, in case of major disaster, area or national representatives may enter a disaster area and take over from local representatives the direction of certain Red Cross activities[7].[xiv]

MAP #3: Distribution of Security Agencies (Following Page)

---

[7] See American Red Cross, When Disaster Strikes: A Chapter Manual for Disaster Preparedness and Relief (Washington: American National Red Cross, 1948), pp. 1 -7; American Red Cross, Mass Care in Disaster (Washington: American National Red Cross, 1951), pp. 51-53.

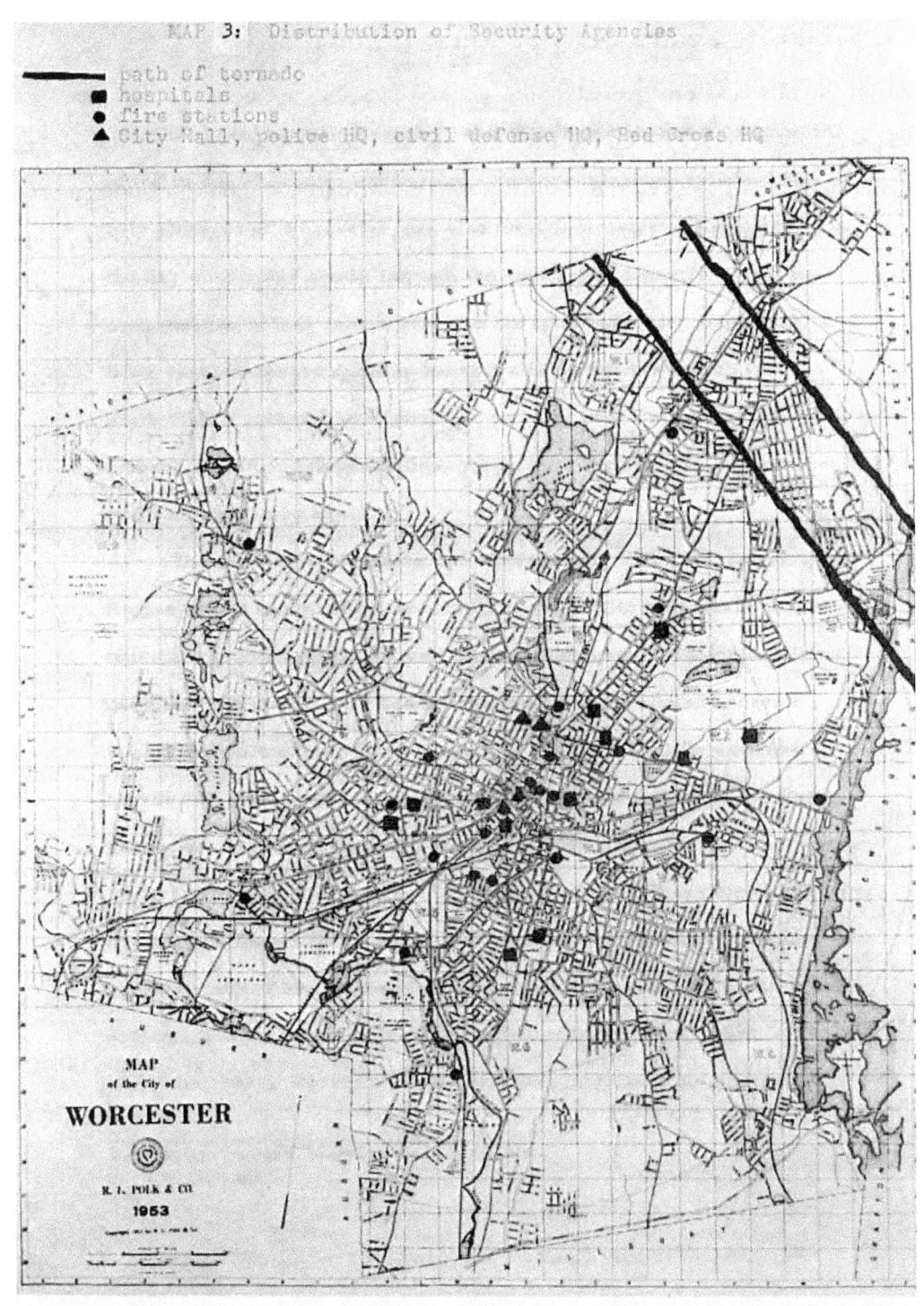

MAP 3: Distribution of Security Agencies
— path of tornado
▪ hospitals
● fire stations
▲ City Hall, police HQ, civil defense HQ, Red Cross HQ

It would be possible to continue at some length to describe policies, organization, and equipment of the couter-measure agencies, including some which I have not mentioned: The National Guard, the Weather Bureau, the insurance network, and so on, but this would require more space than is available. It would also be interesting to study Worcester's experience in the hurricane of September 21, 1938, when seventeen were killed and one hundred sixty-six injured in Worcester County. But this too would be beyond the scope of this report.

II: WARNING: The City Learns of Impending Impact

Worcester lies in the northeastern United States, where tornadoes are believed to be less prevalent than to the south and west. Many Worcester residents, after the disaster, commented on their surprise at a tornado, which they associated with other parts of the south and west[8]. Tornadoes might hit other people; they might even hit Ohio and Michigan (Worcester residents had been reading about the tornadoes at Flint and Cleveland the day before) but not in New England . . . Hurricanes, on the other hand, are (at least after 1938) understandable!

A warning that a tornado was in the area apparently reached the telephone company offices in Worcester before 3:45 PM on June 9th. The telephone company alerted its linemen. The Worcester Telegram city desk got word of this, and the day city editor visited the telephone company office to verify the information. The company admitted alerting the linemen but "refused to divulge the source of their information." The city editor then called the Associated Press in Boston and had them check with the weather bureau. The weather bureau said they had no tornado reports, only reports of severe storms expected[9].[xv]

A similar account of an abortive warning, which was not widely transmitted to the proper agencies, involved the military air base nearby. Weather reconnaissance planes are said to have spotted the tornado early in the afternoon, reported it, and returned to the base, which undertook its own storm-protective measures[10].

A third warning was said to have been contained in a weather broadcast during the afternoon from an unidentified station. This broadcast reported a tornado near Pittsfield, Massachusetts (south-west of Worcester, near the New York State border). A severe storm did actually occur near Pittsfield, but no tornado, although a funnel-shaped cloud was reported nearby[11].

Within minutes after impact at Worcester, however, a radio warning did go out over a police network which was picked up by a patrolman in Fayville,[xvi] where the impact occurred at about 5:30. The patrolman drove home, to protect his two sisters and their four children, but impact caught them sitting in the living room and they never reached the cellar[12]!

MAP #4: The Path of the Tornado – Following Page:

---

[8] I suspect that the assumed relative immunity to tornadoes by the general population in the northeast is a cultural phenomenon. The Iroquois Indians of New York State were so much impressed by tornadoes that the whirlwind spirit (False Face) was given a central place in ritual and belief. Flora, 1953, indicates that tornadoes are not infrequent in the northeast, although some other regions have more.

[9] Powell, MS 1953

[10] Field notes.

[11] See Worcester Telegram, 10 June 1953, p. 17. These reports on warnings which were not transmitted or taken seriously are, for obvious reasons, difficult to verify (because if they are true, they imply either negligence or incompetence in some one). The reports I note above are not verified but they may be substantially true. It might be speculated that this sort of complaining we-could-have-been-warned and why-weren't-we-warned rumor may often rise after disasters; certainly such stories acquire considerable publicity – e.g., the report of the ignored radar operator's report on unidentified aircraft just before the Pearl Harbor attack.

[12] Worcester Telegram, 10 June 1953, p. 11.

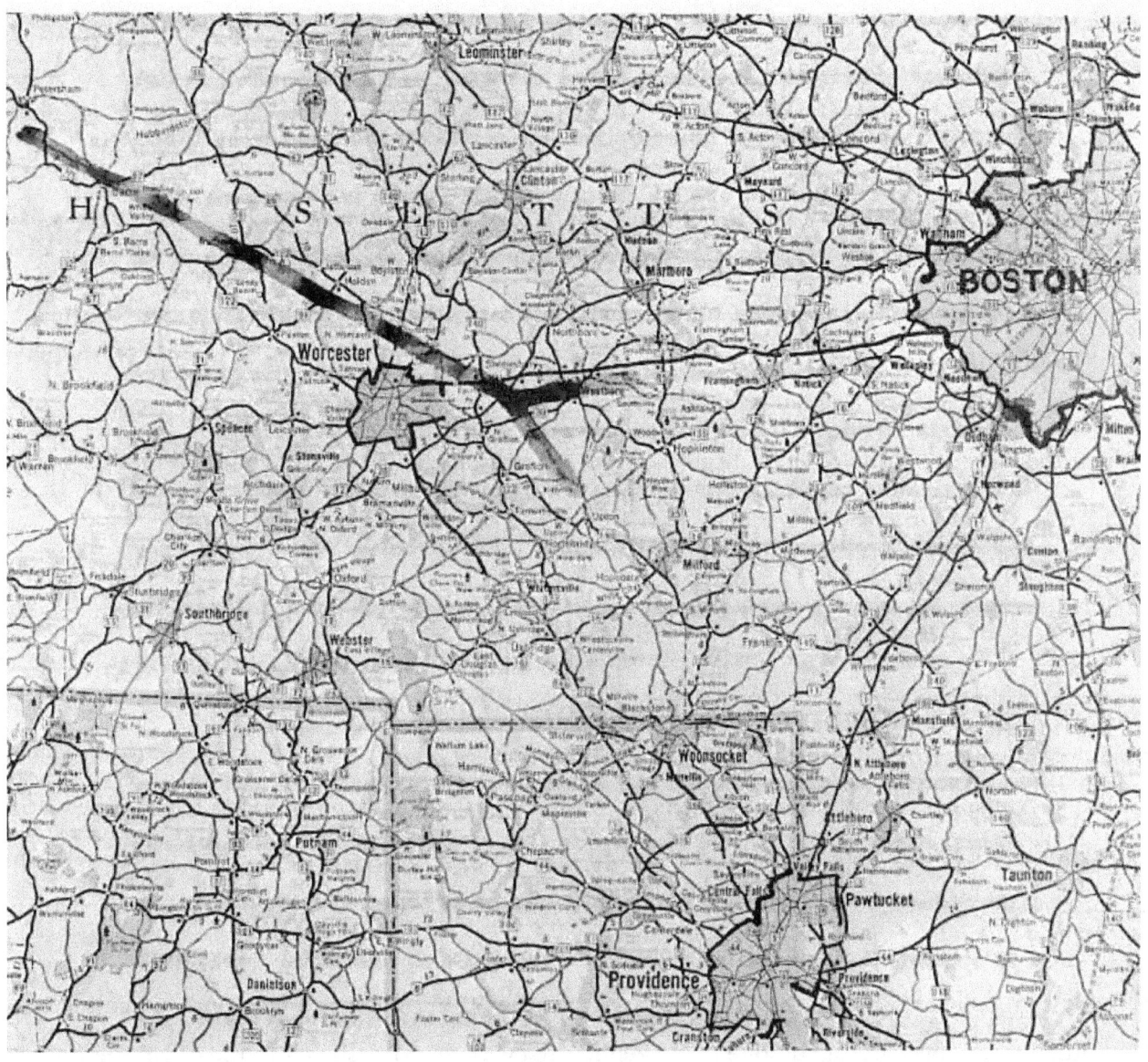

The tornado itself, began its career in the Worcester area at Petersham, about twenty-five miles north-west of Worcester, at approximately 4:30 PM. It was not until about forty minutes later that it reached the city limits of Worcester; it was seen at 5:08 PM at approximately the northern city line. Theoretically, therefore, if a tornado had been reported to weather bureau at 4:30, even though its future course would be unknown to the observer, a tornado warning could have been broadcast to all surrounding communities. Apparently this was not done. Failure to do so was the subject of complaint by one informant, a representative of a radio station, who said that there were newsmen at Petersham who could have (assuming telephone circuits were open) at least called the station office in Worcester[13]. Thus, although the telephone company and the nearby airbase may have had warning as much as an hour and a half in advance, and may have taken local security measures, the

---

[13] Interview with Adolphus J. Brissette, 16 June 1953.

community at large had no warning until the tornado was a few minutes away from successive points in the eventual impact area.[xvii]

The next cue – but one rarely, if ever, interpreted as a tornado signal by the Worcester residents – was the appearance of enormous hailstones. These began to fall about 4:45 PM in the western fringe of the Worcester area; one caller reported a stone eleven inches in diameter. Although this type of enormous hailstone is diagnostic of a nearby tornado condition, Worcester residents did not know what they meant, and no one is reported to have taken them as tornado warning signals, although one informant after another reported noticing them and marveling at their size[14].

The photograph taken at 5:08 PM by a news photographer standing on the east shore of Indian Lake shows the tornado-cone, whirling debris, heavy clouds, and lightning – at a point about two miles distant, near the city line[15].

There were a number of persons in and near the impact area who saw the cone, recognized it as a tornado, and took some protective action. Since the tornado as a whole was moving at approximately a half-mile per minute, this meant that if the tornado were seen at a distance of two miles, the spectator had about four minutes to take cover or get out of its path. It is impossible to estimate how many persons within the impact area saw and recognized the tornado, but probably there were several hundred. [xviii]

The potential grace-period afforded by sight of the funnel, ranging up to four or five minutes, was used by those who did see and recognize it primarily for warning others and taking cover. Thus, a man in Curtis Apartments saw the funnel in the distance, recognized it, ran up and down the halls of his wing warning the other residents, and succeeded in getting most or all of the families down into the cellar before impact. A fourteen-year old boy, delivering papers in the impact area, saw the funnel, ran into a nearby house and told the people a tornado was approaching; despite their skepticism, everyone took shelter in the cellar; the house was

---

[14] See Powell MS, 1953.
[15] Cf. Tornado, p. 2; WDT, 10 June 1953, p. 22 for photograph.

Plate 1: THE TORNADO CLOUD AT TWO MILES DISTANCE
This picture was taken at 5:08 looking northwest from a position on the east shore of Indian Lake.

destroyed.[16] "Scores" of persons in the tornado's path were reported by the newspaper to have taken to the cellar after seeing the funnel, and to have remained in safety there while the superstructure of the house was damaged or destroyed[17].

Probably the great majority of the population of the impact area did not see the funnel, and, consequently, did not know that a tornado was approaching them. The experience of these people, while various in detail, was generally one of considering the darkening of the sky, the rising wind, and the pelting rain to be a summer thunderstorm. Those who were out of doors ran into the houses; those in the houses ran to close doors and windows. Many of these did not realize that this storm was "different" until a few seconds before maximum impact; at this point, when the scream of the tornado (generally described as sounding like a locomotive or a jet plane) was audible, the air was beginning to fill with mud and debris, and windows and trees were beginning to break, an almost intuitive awareness that disaster was upon them tripped off emergency action even without rational understanding of the event. These people had only a few seconds before maximum impact. Many tried to reach the cellars (often from the second or third floor of houses); many, for some reason, tried to open doors and get out of the house; some took cover under furniture, in closets, under the stairs, etc.; many began a frantic search for children or relatives. In many cases, these actions were uncompleted at the time of impact; the cellar had not been reached, children had not been gathered together, the door wouldn't open (fortunately).

An intensive analysis of this last-second-reaction period would be interesting to FCDA, since it approximates the situation envisaged in CD popular instructions on "what to do when there is a blinding flash."

I have made an analysis of fifty more or less adequate case histories of persons in the Worcester impact area at the time of the tornado. The total population of the impact area is uncertain, although it could be computed approximately; it probably was about 9,000 persons. The fifty interviews are an undefined sample of this universe of impact-area people. The interviews themselves used in the series of fifty were obtained from several sources: Jeannette Rayner, an experienced disaster interviewer with a psychiatric orientation, three interviews; Edna Barrabee, a psychiatric social worker, nineteen interviews; John Powell, an experienced disaster interviewer with a psychiatric orientation, three interviews; Enoch Callaway, working for Powell, three interviews; Anthony F. C. Wallace, two interviews; obtained by reporters on the Worcester Telegram and Gazette, sixteen interviews; obtained by reporters of Radio Station WTAG, four interviews. This sample tends to be heavily weighted in favor of persons severely injured and consequently available at the hospital to Rayner and Barrabee, who were doing a study of medical care; and persons with minimal physical injury, but with a

---

[16] Powell, MS 1953
[17] WDT, 10 June 1953, p. 22

dramatic human interest story to tell to newsmen. Nevertheless, granting the uneven quality of the interviews and sampling inadequacies, the high or low frequencies of certain characteristics in this series are suggestive of at least similarly high or low frequencies in the impact population.

The distribution of interviews by locality of interviews at time of impact is as follows:

TABLE 2: DISTRIBUTION OF INTERVIEWEES: BY LOCALITY (see map)	
Brattle Street Area	1
Industrial Area	1
Greendale Area	14
Assumption College	3
Burncoat Street Area	1
St. Nicholas Development Area	5
Great Brook Valley Area	11
Curtis Apartments	6
Lincolnwood	2
Home Farm	5
Not Recorded	1
TOTAL # =	50

While this probably is not a proportionate sample, there is at least a considerable scattering of interviewee locations. The composition of the series by sex and injury is instructive; there were twenty-eight males and twenty-two females; twenty-five seriously injured and twenty-five slightly injured or uninjured. The women in the impact area tended to be more seriously injured than the men: only eleven out of twenty-eight males were seriously hurt, but fourteen out of only twenty-two females were seriously hurt.

Not one of the fifty cases had received any official warning (i.e., by radio or telephone, police loudspeaker, etc.) of a tornado's being in the region. Thus, warning possibilities resolved into a recognition of any one of a series of cues, beginning with the fall of unusually large hailstones (a frequent but not invariable omen of a tornado), and culminating in the approach of a visible funnel, increasing wind, loud noise, and, finally, personal impact.

Not one of the fifty indicated that he recognized the hailstones as possible tornado indicators, although many remembered having seen them and being impressed by their size as long as fifteen minutes or more before impact.

Twenty-two out of the fifty saw the funnel of the approaching tornado; but only fourteen of these twenty-two recognized it for what it was, as indicated below:

TABLE 3: INTERPRETATIONS OF SIGHT OF FUNNEL
Recognized it as tornado = **14**
Did not know what it was = **5**
Considered the possibility of it's being a tornado, but dismissed the thought = **2**
Thought it was smoke from a fire set by lightning in thunderstorm = **1**

Thus, of our sample of fifty cases, only fourteen – less than one-third – saw and recognized the tornado cloud. The data recorded do not give more than hints as to factors responsible for not seeing the funnel: a few people were asleep, some were in a house or apartment with windows facing to leeward, and some apparently could have seen it but just didn't notice (automobile drivers in the area, children playing, etc.).

Of our fifty cases, thirty-nine (78 per cent) recognized some warning cue before personal impact.

Fourteen (28 per cent of the total) saw the funnel and recognized the danger; twenty-one (42 per cent of the total) recognized a danger from other clues closer to impact time; only four (8 per cent of the total) were warned by others.

Of our fifty cases, twenty-eight (56 per cent) were able to complete some personal protective measures before personal impact. Twenty of these twenty-eight (40 per cent of the total) reported warning others or helping others in protective action before completing their own measures; only eight (16 per cent) reported taking protective action oriented only towards their own safety[18]. Twenty-two persons (44 per cent of the sample) were unable to complete (and in some cases, unable even to initiate) protective action before personal impact.

Of these twenty-two, six suffered impact while attempting tardy protective action, but five took no protective action even during the impact itself.

---

[18] N. B.: We are depending on reports of own actions.

III: IMPACT: The Tornado Strikes

At approximately 5:08 PM, EDT, the tornado reached the Worcester city line. It had been traveling in a southeasterly direction, and (unbeknownst, apparently, to anyone in Worcester) had originated at Petersham, near the Quabbin reservoir, about forty minutes earlier, at 4:30 PM. The tornado, as a body, was moving forward at about twenty-five miles per hour, and crossed the northeastern corner of Worcester (a path three and a half miles long) in almost exactly eight minutes[19].

The diameter of the vortex itself was approximately a half-mile. Since the forward motion of the tornado was at a rate of a little less than half a mile per minute, the passage of the vortex over a point in the middle of the path required little over one minute.

This was an unusually severe tornado: it was large in diameter; the velocity of the wind in the outer ring was extremely high (estimates of winds in comparable tornadoes go as high as 500 miles per hour); and the air pressure within the high velocity ring was very low. No reading of the pressure within the vortex was made, but in comparable tornadoes where readings were made, the pressure dropped as much as two pounds per square inch. The funnel was photographed (see Figure 1) at a point near the Worcester-Holden line: it shows a broad and high, roughly funnel-shaped cloud, but does not show a clearly outlined funnel touching the ground. This was probably hidden behind trees and a hill. Several informants reported seeing a vortex shaped like a "snake," an "ice cream cone in the sky," or "a whirling cloud," but this may have been a secondary elaboration in fantasy, since a base a half-mile in diameter, spewing out mud and debris, would even at a distance of one or two miles probably not appear to have a very sharp outline. The color was reported as "black," at a distance of a mile or two, but "brown" (with mud) at close range, and circulating boards, boxes, paper, and other debris were clearly visible in it. It made a very loud sound, described by survivors as reminding them of several steam or diesel locomotives, or a flight of jet planes as it approached; at impact, the pitch rose to a "scream."

The path of the tornado through Massachusetts and through Worcester, where it maintained continuous contact with the ground, is charted on Maps 2 and 4.

The primary impact of the tornado is the physical damage and injury wrought by the tornado during its eight-minute passage through the impact area; certain other sorts of hazard or disaster, such as fires, electrocution by "hot" (live) wires, sepsis in untreated wounds, and consequences of destruction of food, clothing, and shelter, are classified as secondary impact and will be treated later. The primary impact may be conceived of as a three-stage process for most of the impact area: high wind, "vacuum," high wind again.

---

[19] WDT, 10 June 1953, p. 22.

MAP 5: Disaster Space at I-Time (5:08 to 5:20 P.M.)

• • • • • area of total impact
━━━━━ area of fringe impact
• ━ • ━ filter area
The other areas have not yet been defined.

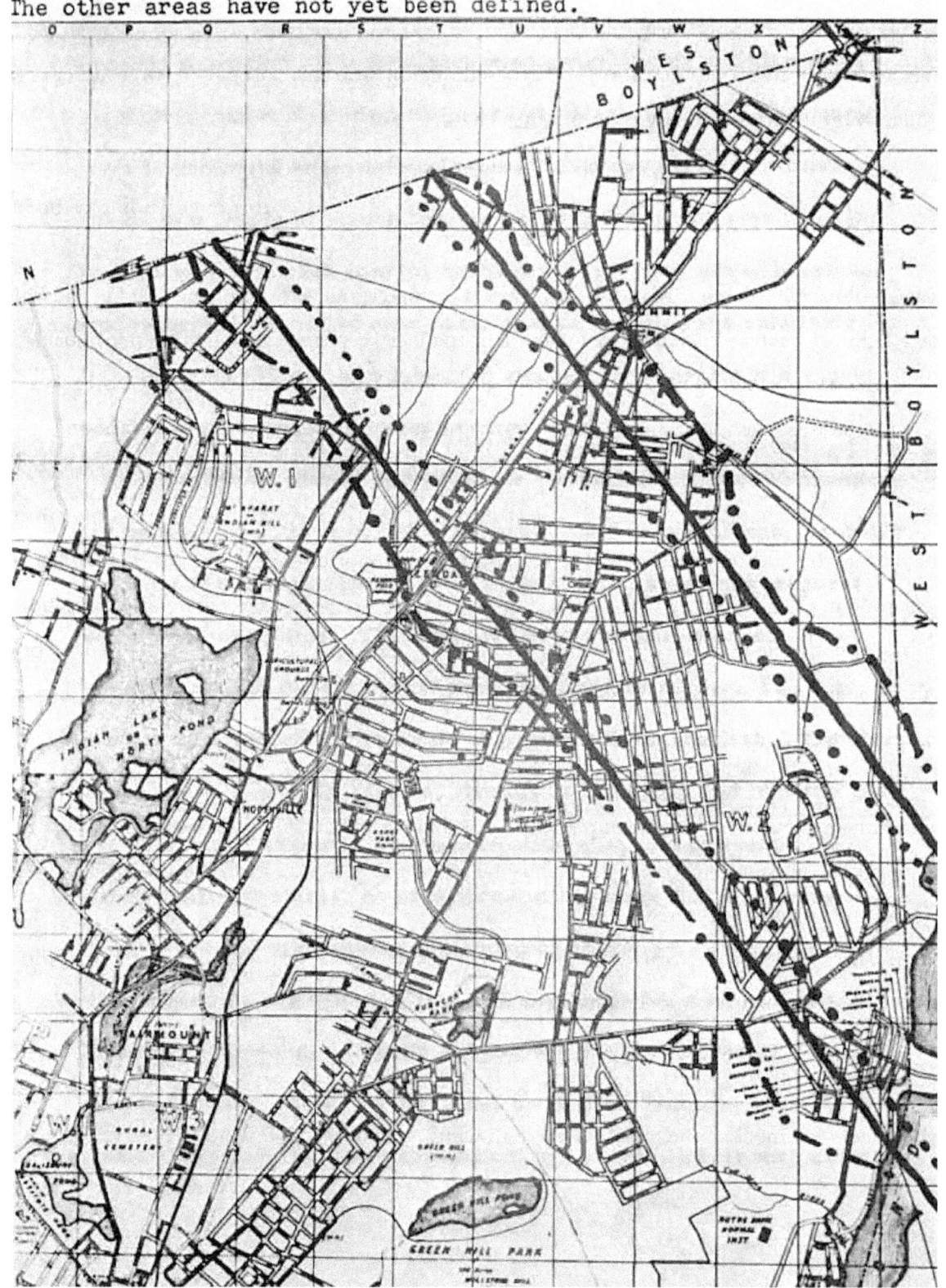

The first stage is high velocity wind as the front of the vortex crosses a position. The first high wind generally did not blow down structures; but it blew windows in or out, depending on whether they were to windward or leeward; plastered buildings with mud; drove debris into people or structures, sometimes driving objects through wooden walls; knocked down or bent over trees, flag poles, hurricane fence; toppled and rolled over cars, shacks, people, and relatively light movable objects; and subjected structures generally to stresses which tended to weaken or displace members.

The second stage – low pressure – apparently was not accompanied by high velocity wind. This however, was, to structures, probably the most destructive aspect of the whole impact; survivors' reports seem to agree on this. The first high wind suddenly stopped and at almost the same instant the air pressure dropped perhaps 10 or 15 per cent, equivalent to two or three pounds per square inch. The quickness of this drop in pressure apparently is dramatic; and it is the sudden drop which seems to account for many of the freak events. In lightly built structures, or structures with a large flat area without "escape valves" like windows, chimneys, doors, etc., the pressure drop meant the equivalent of a sudden application of tremendous forces outward and upward. Actually, of course, the pressure within a structure does not increase; but the withdrawal of air from the outside meant that inside air was pressing out against all objects with about fourteen pounds to the square inch, while the outside air was pushing back with perhaps twelve pounds per square inch: a differential of two pounds. Applied for example to an outside wall of a house eight feet high and ten feet long, there was thus an increase in outward lateral thrust from zero to eleven and a-half-tons within a period of a few seconds at most; and this occurred immediately following the application of considerable stress, in many instances in an opposite direction, from high velocity wind. Many structures simply exploded: informant after informant describes how, after a few moments of raging wind, the whole building seemed to dissolve, in slow motion, and they would find themselves sitting in quietness and relatively uninjured in the yard, surrounded by pieces of house. Where the structure did not "explode," the rush of air from inside proceeded to blow out windows and doors, and to carry out any movable object – including people, refrigerators, television sets, chairs, and tables. This "floating away" was made possible, apparently, by two forces: (1) the lateral evacuation of air through vents like broken windows and doors; and (2) by a vertical differential air pressure, which I think must be postulated to explain the lifting of objects and the frequently referred to "floating" phenomenon. An example of this "floating" phenomenon, which baffled victims and added to the uncanniness of the experience, is given by the little girl who was in the kitchen with her mother and father at impact. The mother and daughter were cooking supper, and disregarded the first high wind as being only a bad thunder storm. They had just put the potatoes for baking in the oven when the pressure differential came. A strange thing happened: "the potatoes came out of the oven and went over and hit my daddy on the head." A woman saw a pane of glass, blown in by the high wind, float gently to the cement floor of her cellar without breaking. Two mothers reported seeing their children float away from their side; each time the mother grabbed the child and pulled it down, like pulling in a balloon floating away. Many observers reported seeing heavy objects float across the floor toward a window, not scraping the floor, and moving slowly. One man. was carrying a crate of eggs; "the eggs were popping out of the crate but they weren't falling to the ground." The explanation of the floating phenomenon must be that when a "vacuum cap" is placed over the area, the "free" air out-of-doors is sucked out almost instantly; the air within structures goes next; but in-structure pockets of high pressure air remain, particularly under objects like chairs, TV sets, and even people, and in places like ovens, egg crate partitions, etc. It is quite obvious, for instance, that if even so small and dense an object as an egg or a potato or a child's body were placed between a region of twelve-pound air above and fourteen-pound air below, the pressure differential on the upper and lower surfaces would be sufficient to float it. The pressure differential may help explain how cars, locomotives, and even sidewalks are lifted up. Inasmuch as this floating phenomenon was very commonly reported, it would seem reasonable to suggest that in any type of disaster, like tornadoes, where vacuum effects are common, the population be prepared to cope with it.

It is my impression (I cannot document it adequately) that it was the explosion effect which actually demolished most of the structures which were demolished, but that it was the severe high wind, following the explosion, which caused a larger share of the severe casualties and deaths. Except for those who were caught in the open or were in vehicles at the first wind, most of the interviewees weathered it in their own structures. The "explosion" was tempered by the floating effect, which tended to lift the roof off, and push the sides out, rather slowly, so that many persons were either not covered by the debris, or were not crushed to death when it did come down on them. But after perhaps fifteen to thirty seconds of the "vacuum" the high wind came back (the back wall of the tornado). This caught many people in a bad situation: many were now in the open; they were dazed and bewildered; they were surrounded by loose and movable debris. The second passage of high velocity wind thus, so to speak, "finished the job," by blowing survivors about, or by blowing debris onto or into them, or both. Some people also ran out of relatively undamaged structures during the vacuum period, and were then swept up by the second, high wind.

Thus for the person in the central path, there was a triple 1-2-3 punch to weather: an initial high wind with dangers of severe injury or death if caught outside, or of breaking glass and structural weakening if inside; then the vacuum, with the "floating" effect and slow "explosion" of structures; and finally, the second, high wind, which struck structures which had either crumbled or were severely weakened, and found many people in the open surrounded by the ruins of their shelter.

This generalization, of course, applies best to those people and structures dead center in the tornado's path. For a person at the extreme edge on either side, there was presumably no "vacuum" period and only one long high wind period. Intermediate points between edge and center no doubt experienced varying lengths of vacuum period, and directions and lengths of high wind.

The impact area can be divided into an area of "total" impact, where structural damage was severe and the population suffered deaths and many severe injuries, and an area of "fringe" impact, where structural damage was slight (broken windows, torn shingles, mud-spattering, broken tree limbs, and the like). The total impact area itself can for the sake of convenience of reference and analysis be further subdivided into ten neighborhoods, which are located and numbered on Map 6:

1) Brattle Street area. A middle class residential neighborhood including large frame detached single homes in the $20,000 – $30,000 class, on large lots, on high ground, as well as smaller bungalow-type housing, looking out over undeveloped land.

2) Industrial area. The Norton Company (specializing in grinding and abrasives equipment); Vellumoid Corporation (makers of gaskets and gasket materials); the Diamond Match Company; and other manufacturing and business concerns, and some deteriorated housing, were located here. The Industrial area is in a valley.

3) Greendale. A lower-middle class to middle class residential neighborhood, mostly frame structures, including various small shops and stores, some detached homes, some "3-deckers" (each story serving as a "tenement" or apartment). Three churches were located in this area, and it seemed to house a number of elderly or retired persons who had invested their life savings in a house (often a "three-decker") of which they occupied one floor while renting out the rest. Greendale is in a valley and on the slope of the rather steep hill that rises to Burncoat.

4) Assumption College. A Catholic college and secondary school conducted by lay brothers and priests of the Assumptionist order. Many of the staff of priests, nuns, and brothers were French-Canadian in origin, and many spoke only French. The plant included a brick main administration building and a wooden convent. Assumption College's structures are all in the valley, although the College land reaches up to Burncoat Street.

5) Burncoat Street area. Upper Burncoat Street is a "nice" residential area on the top of the hill with a small shopping center, houses ranging from perhaps $15,000 to $30,000 in value, relatively large lots. St. Michael's on the Heights Episcopal Church is on Burncoat Street.

6) St. Nicholas Development area. This is an area of post-war development housing, of two major types: two-story, four-apartment frame structures for rent; and single story detached bungalows on cement slabs. There are, however, other types of detached single-family houses within this area. The families here tend to be a typical "development population": young couples, small children, careers in the making.

7) Great Brook Valley. This contained two types of housing: some privately owned, detached, single-family frame dwellings; and Great Brook Valley Gardens, a housing project, owned by the Worcester Housing Authority, consisting of one-story brick row houses, with picture windows, completed just a few weeks before impact.

8) Curtis Apartments. These were large brick apartment buildings (also owned by the Worcester Housing Authority). 3000 people lived here.

9) The Home Farm. A city-supported home for old and indigent persons. Administration building of brick; wooden barracks; farm buildings.

10) Lincolnwood. War temporary housing, consisting of twenty-two reconverted frame barracks. Many of these had been removed, and many were standing empty. Some scattered privately owned dwellings stood here too.

MAP 6: Neighborhoods in the Impact Area

1. Brattle Street
2. Industrial
3. Greendale
4. Assumption College
5. Burncoat Street
6. St. Nicholas Development
7. Great Brook Valley
8. Curtis Apartments
9. Home Farm
10. Lincolnwood

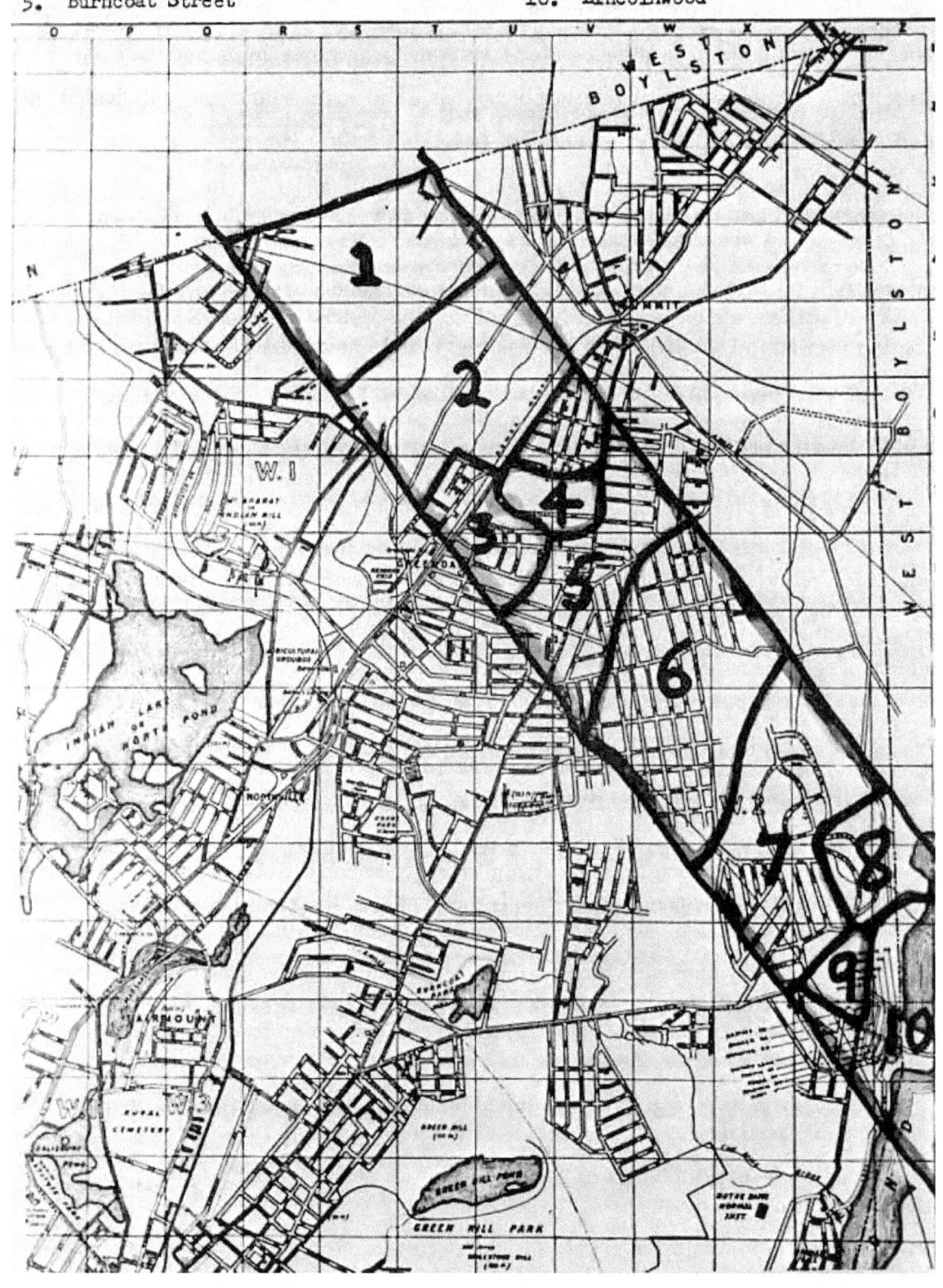

The total amount of physical trauma produced by the primary impact was very great. The impact area within the Worcester city limits was approximately two square miles. Within this area were about 1800 dwellings, the homes of about 9,000 persons. Since a large part of the labor force component of this population was outside of the impact area (coming home from work) at i-time, as well as some mothers et al who were shopping, visiting, etc., probably the population of the impact area at i-time was not more than 8,000 and possibly was much less. Of this population, 804 (about 10 per cent) were "casualties" (here defined as being killed or as being taken to the hospital). Of the 804, sixty-six were killed (less than 1 per cent of the impact area population); 327 suffered major injuries; and 411 suffered "minor" injuries. Presumably all deaths and major injuries are included in these figures, collected by the Red Cross from hospital records, but no doubt there were many minor injuries which were either left untreated, or were treated privately by physicians, nurses, neighbors, or the victims themselves.

Estimates of these would be very difficult to verify; as a guess, one might suggest that if these unreported minor injuries (slight cuts, bruises, scrapes, etc.) were included, the total casualty rate would be about 25 per cent of the impact area population[20].

An analysis of fifty-six of the sixty-six deaths by sex, age, and residence indicates chiefly that (as one might expect from knowing that many wage-earners were outside of the area at i-time) females suffered more than males:

TABLE 4: FATAL CASUALTIES BY NEIGHBORHOOD	
(1) Brattle Street Area	1
(2) Industrial Area	0
(3) Assumption College Area	3
(4) Greendale Area	9
(5) Burncoat Area	5
(6) St. Nicholas Area	8
(7) Great Brook Valley Area	15
(8) Curtis Apartments Area	0
(9) Home Farm Area	5
(10) Lincolnwood Area	0
Visiting in or driving through impact area (or resident elsewhere in Worcester or Worcester County)	8
Total N =	54

The data available do not make possible a tabulation of specific cause of death or of the behavioral circumstances at the time of fatal injury, but according to the medical report, most of the deaths were caused by or associated with severe cranial injuries.[21]

---

[20] Basket et al., 1953
[21] Basket et al., 1953 Appendix

Several casualties died of lung injuries resulting from the "vacuum" condition and several informants remarked on having difficulty breathing during impact. Some bodies were found with an empty cranium; presumably the cranial contents were "sucked" out of the cranial cavity after the cranium was crushed. [xix]

TABLE 5: FATAL CASUALTIES BY AGE AND SEX

Age Group	Male	Female
0 > 9	3	0
10 > 19	4	3
20 > 29	4	0
30 > 39	2	4
40 > 49	2	2
50 > 59	3	2
60 > 69	3	1
70 > 79	4	0
80 > 89	1	0
Unknown	7	11
Totals	22	24

The Red Cross tabulated the type of injury in the 438 "major injury" patients, as follows:

TABLE 6: TYPES OF MAJOR INJURY	
Fractures	242
Skull (usually involving cerebral injury)	77
Arms and Legs, Hands, Feet	88
Ribs	32
Shoulder	21
Pelvis, Hip	15
Misc. (no broken backs)	9
Soft Tissue	210
Major cuts and bruises	147
Eye	28
Back	15
Internal (Kidney and/or Spleen)	15
Burns	5
Total Major Injuries	452

It was also noteworthy that most casualties (and, indeed, any person exposed to the high winds) had been literally plastered with a mixture of water, dirt, sand, wood splinters, etc.; where there were open wounds, this foreign matter had been driven into the wounds with such force that adequate cleaning was very difficult, and a number of cases of infected wounds turned up later on (after suturing).

The statistics of structural damage in Worcester are only approximate. About 2500 dwelling units were in the impact area. Of these, 250 were totally destroyed; about 1200 seriously damaged, and about 1000 were slightly damaged. The financial cost of the destruction in Worcester alone was about $32,000,000[22].

In the entire disaster area, from Petersham to Wrentham, the total number of dead was 94, seriously injured, 438; about 750 minor injuries were hospitalized the night of June 9; about 4000 dwelling units were damaged or destroyed; 1200 persons were unhoused; total damage was in the neighborhood of $52,000,000[23].

(See the pages of photographs for a survey of the physical damage in Worcester, area by area.)

The sixty-seconds or less of the impact period were, for most of the people in the impact area, a time of busy physical activity. On forty-nine of the fifty survivors whose behavior during the warning period was surveyed, there is also information on what they were doing during impact (i.e., during the whole of the passage of the vortex, or during the part of it that preceded disabling personal injury). Sixteen of these forty-nine (about one-third of the group) spent the i-time finding and enjoying shelter: six lying on the open floor of a room, five in the cellar, three in a closet or other enclosed space, and two under furniture. Seven were inside a structure and running to a place of shelter (such as the cellar or a closet when they were struck personally and severely injured. Seven were attempting to manipulate (usually to close) doors and windows when they were struck and severely injured. Six were rushing about inside a structure, trying to care for other persons, when they were struck. Two (both clergymen, incidentally) stood in doorways gazing at the storm all during i-time. One person grabbed a pillar for support. Only two persons were apparently unaware of the tornado until personal impact; both were women; and one was sitting on a porch talking about the weather, while the other was washing her hands in the kitchen; neither of these attempted any protective action before or during the impact. Six persons were in cars during i-time: four of these were able to crouch under the dashboard; one was trying to stop the car when the car was picked up and blown into a field; and one was able to drive out of the tornado's path. Two persons on foot in the open were running to shelter when they were injured: one was blown against a tree, and one was knocked down by a collapsing brick wall.

---

[22] Bakst: Landstreet; Bowman: Powell; WTG and WTAG.
[23] Ibid.

Plate 2: DEVASTATION AT ASSUMPTION COLLEGE

Plate 3: DEVASTATION IN ST. NICHOLAS DEVELOPMENT AREA

Plate 4: DEVASTATION IN GREAT BROOK VALLEY AND CURTIS APARTMENTS

IV. ISOLATION: The Impact Area Goes It Alone

The period of functional isolation for the impact area varied from a minute or two, in some of the southern fringe areas, to perhaps half an hour in parts of the central and north-western sections of the impact zone. Since the isolation period begins at the moment when the primary impact ceases, its clock-time varies depending on location; the isolation period at Brattle Street was seven or eight minutes old by the time impact had begun at the Home Farm.

The situation at the beginning of the isolation period is the immediate legacy of the impact: a definite quantity of physical damage, personal injury, and death. This primary damage is for practical purposes irreversible during the isolation period; and the energies of the survivors are, during the period before the arrival of aid, necessarily directed toward preventing further death, injury, and destruction, from the processes set in motion by the primary impact.

These potentially destructive processes, launched by the primary impact, include such things as:

    a) Bleeding from wounds
    b) Interferences with respiratory processes
    c) Wound shock
    d) Development of various "infections" in wounds
    e) Entrapment
    f) Fires, set by lightning, cigarettes, short-circuiting wires, sparks, etc., and feeding especially on spilled gasoline and escaping cooking gas;
    g) "Hot" high voltage or heavy current wires lying in the way of moving persons
    h) Exposure to rain, sun, cold, etc.
    i) If isolation is prolonged, hunger and thirst.

These potentially destructive processes, set in motion by the primary impact, may collectively be called secondary impact. During the time before effective assistance, particularly organized assistance, by police, fire, civil defense, medical, and other organizations reaches a survivor, he must cope with secondary impact alone.

The length of the isolation period varied in Worcester. Organized aid in the form of police cruiser cars, fire equipment, and ambulances, reached the Home Farm, Great Brook Valley, and Curtis Apartments areas within five to ten minutes. This was owing in part to the fortunate accident that telephone communication with central Worcester had not been cut off by primary impact. Within about a minute at least three telephone notifications to Worcester security units had been made: the assistant executive director of the Curtis Apartments, himself a minor casualty (his car was blown about and he was cut by glass), heard a phone ringing and used it to call police, fire, City Hospital, gas company and electric power company; a housewife in Great Brook Valley used her own phone to call City Hospital and tell them to send ambulances; and someone (unidentified) told the Fire Department headquarters that there had been a boiler explosion at the Home Farm. A box alarm was turned in at the Home Farm at the same time. A ladder company with rescue crew and first aid equipment was on the way toward the Home Farm at 5:15 PM; this company, indeed, nearly ran into the tornado itself on the way down Plantation Street, and managed to get into the hail zone on its fringes. The path of the tornado and of the fire engine were approaching an intersection south of the Home Farm, and the tornado got there first by a minute or two.

This ladder company set to rescue work at the Home Farm within five minutes of impact; and they were helped in driving into the area by the crews of police cruisers who were already setting up a road block on Lake Street.

It is difficult to pick out the particular area which was last to be taken out of isolation by the arrival of aid, but probably portions of the Burncoat Street and St. Nicholas Development areas, and the Brattle Street area (difficult of access on account of narrow streets, hills, trees, and traffic jams) were last to be invaded by rescue forces. But even here the isolation period was not long; probably no point in the impact area was left un-checked, or had been unable to communicate with sources of aid, by thirty minutes after impact. The average length of the isolation period was probably about fifteen minutes; by that time residents of the filter area, firemen, or police had appeared and were within calling distance.

The general scene during this short isolation period is difficult to reconstruct because outside observers had not arrived and those in the area were not always in the best position or frame of mind to observe objectively (or to remember without distortion later). A generalized reconstruction from the comments made by survivors, however, suggests a scene like this:

While many persons had been taking cover in cellars or closets, many others had been frantically active all during the impact. These activities carried on directly into the isolation period, with searching for children, scurrying for cover, etc., continuing until it was obvious that the primary impact was really past. Then ensued a brief frenzy of action and reaction: sudden frantic efforts to rescue self and trapped relatives, screams and cries for help, hysterical laughing and crying, particularly (as reported) by teen-age girls, people rushing up and down stairs, into and out of cars, houses, etc., checking on the welfare of others; shouted warnings to "look out for the wire," "don't light any matches," etc. Some, able to walk, spontaneously ran or were dispatched to neighboring fringe and filter areas to summon aid: these persons turned in fire alarms at boxes, called security agencies (police, fire, and hospitals) by phone, and brought in trucks, people with bandages and antiseptics, and other first aid resources. Other persons were busy in self or local rescue, usually of a relative living in the same or adjacent house. Even severely injured persons made efforts to extricate themselves, to summon help, to estimate the seriousness and character of their own injuries, and to give instructions to rescuers on what to do and not to do. The reports which I have seen do not show that impact was followed – as is sometimes asserted – by a deadly calm, silence, and absence of motion. Quite the contrary: there was a great deal of noise and much self-help and mutual aid activity.

For those whose injuries were severe but who were conscious the general pattern of action was:

(1) Immediately after impact, to orient themselves (by finding out where they were physically, what their injuries were, and what should be done to get help, or to rescue themselves);
(2) To extricate themselves physically, if possible, or if that is impossible, to call for help;
(3) If extrication was achieved, to go towards help, or to a protected or more comfortable place;
(4) Occasionally, to render some assistance or give some directions to other members of the family;
(5) Once having oriented themselves, achieved rescue or summoned help, and done what was possible to help or direct other family members, to subside into a dazed and apparently apathetic state.

For those who were not seriously injured, the general pattern was:

(1) Personal orientation (which was often less of a task, since many of these people were in cellars, closets, etc.);
(2) Personal extrication and minor first aid if necessary;
(3) Rescue and first aid to personal family members (if any);
(4) Awareness of the extent of community damage.

The uninjured group, however, splits at this point into a group who, having cared for self and family, do nothing more to aid neighbors or the community, and those who proceed to rescue, communications, first aid, etc. Among the former (non-helpers) are some who:

(5) Simply block out of emotionally significant awareness their perception of the extent of damage to the community and the possibility of their being of help. (One of the cases interviewed was not aware for perhaps half an hour that anyone else had been affected but himself.)
(6) After performing minimal care for relatives, do nothing at all (simply sit or stand or wander aimlessly) or do inconsequential tasks like searching through clothes, papers, etc.
(7) After caring for relatives, simply play a sightseer's role (in one case complaining later that he would have been too much upset by the sight of bloody, mangled bodies to take part in rescue work).

It seems, in the group of non-helpers, that the transition from a family-oriented rescue role to a community-oriented-rescue role is blocked approximately at the time when such a person, having cared for minimal, immediate personal and family needs, first recognizes that the visible community itself has been smashed. One might speculate that such persons have rather flimsy identifications with community-service roles and that while they can perform these roles if they are supported by visible and tangible evidence of a strong community, when the evidence shows a crushed and maimed community, they are simply incapable of such action as long as they are on their own (although once community aid begins to pour in, they can perform community service roles again).

The helper group, on the other hand, theoretically should have firmly established, deeply internalized community service role-identification systems which operate to govern behavior even in the absence of evidence of external support.

The actions of the helper group, following the first three stages common to the non-seriously injured and uninjured, seem also on the basis of a few cases to follow a fairly regular phase-pattern:

1. Vigorous but rather random rescue, first-aid, and advice activities, with an effort to cover as much area as possible;
2. Awareness of extreme fatigue when community aid arrives; return to care of family;

3. In some cases, a second "go" at aiding the community, followed again by a feeling of exhaustion and futility.

The accounts of survivor-behavior during the isolation period leave the reader with the feeling that it was for the most part limited by the emotional impact as well as by the lack of tools and equipment to a rather narrow range of things: physical extrication of persons; jacking up or propping up unstable debris; setting victims on the lawn, wrapping them with blanket or coat, and offering a cigarette; some inefficient first aid procedures, chiefly bandaging wounds. Little or nothing could be done about live wires, the fires that were starting here and there (at least seven fires were well under way), the leaking of gas or water, etc., beyond mutual warning and efforts to notify people outside. Little could really be done about severe wounds, badly trapped persons, or the securing of shelter from the rain without evacuating the area.

By the end of the isolation period, some people were beginning to do this (chiefly in order to get the wounded to hospitals), but there was <u>during the isolation period</u> no real evacuation movement at all. The reaction-formula seemed to be (the community-oriented persons excepted): to combat secondary impact up to the minimum necessary to preserve life in one's self and relatives, and then, unless specifically asked for aid, to sit and wait for help.

The crucial factor distinguishing the two groups probably is the reaction of the person to the sight of a virtually destroyed community. Again and again in the interviews the phrase "the end of the world" occurs to describe the phantasy of survivors; the sight of block after block of ruined homes, of maimed and bleeding people, fallen trees, scarred and lifeless lawns, bedraggled wires, and everything covered with mud, aroused momentarily in many the thought that this was the earth's last hour, or that an atomic bomb had fallen, or that the whole city of Worcester was in ruins. I would speculate that the "sitting dazed and staring," the "aimless wandering," the "inconsequential small talk," and other irrelevant behavior so commonly reported by the first outsiders to enter the impact area was the behavior of non-helpers who, having done what they had to do for themselves and family, suddenly realized that this was not an accident, this was a community disaster – and simply stopped functioning. The helpers, on the other hand, were able, to incorporate the knowledge into their action-systems that this was not an accident, this was a disaster, and to proceed from self-and-family services to community services.

In a sample of thirty-nine cases (out of the fifty previously chosen) on whom there were data concerning experience during the isolation period, it appears that ten (or 26 per cent) were unable to help either themselves or anyone else. Nine of these cases were immobilized by the severity of their injury or the completeness of their entrapment; one was immobilized by anxiety, although uninjured, and was finally "rescued" after several hours, by an organized rescue team.

Eleven of the sample (or 28 per cent) cared only for themselves during the isolation period: four of these were elderly, single men, residents of the Home Farm, and all were seriously injured persons who rescued themselves; six were housewives, all severely injured and unable to rescue themselves, but active in calling for help and in giving directions to rescuers; and one was a little girl who followed directions and, although out, walked to the ambulance.

- Eight persons (21 per cent) helped themselves and their families but undertook to give no aid to stricken neighbors nor to perform other community aid services. Five of these were men and three were women.
- Four of them were seriously injured.
- Of the uninjured, one was a member of a city security Unit who theoretically should have been doing rescue work, especially since he and his family were unhurt.
- Two other men with no substantial personal or family injury gave no community aid in the disaster area.
- And one woman, a housewife, was preoccupied with locating and maintaining control of her six children.
- Thus, in the sample of thirty-nine, only ten persons (26 per cent) were able to perform rescue, first aid, communications, or other service to the stricken area during the isolation period.
- These ten, however, constituted 67 per cent of the fifteen uninjured among the sample of thirty-nine. Nine were men, one was a woman.
- Five of them had families and all of these five had cared for their families before doing community service.
- Eight of the ten already had a professionally assigned disaster role: three were clergymen, two were firemen, one was a doctor (a psychiatrist), one was the assistant executive director of Curtis Apartments, and one was an auxiliary policeman.
- It is notable that seven of the eight had already chosen these community-responsible roles as full-time careers.
- Two of the "helpers" had had no community responsibility or disaster role at all: one was a housewife in Great Brook Valley, uninjured, who called City Hospital and told them to send ambulances; one was a family man, also uninjured, of Great Brook Valley, who engaged in rescue work until his missing son was brought home, severely injured (he had been caught in the open and blown into a tree).

It is difficult to estimate the significance of the percentages in the various categories above, because the sample is by no means representative: it is proportionately over-represented in seriously injured persons, in persons with professional disaster roles (interviewers turned to them naturally for information), and therefore probably the proportions of "unable to help" and "did community service" are too high.

In summary, it would appear that among survivors without major injury somewhere in the neighborhood of 67 per cent were able within about fifteen minutes after impact to orient themselves, care for family, and begin to help the community as such.

About 33 per cent of these relatively uninjured persons, within this approximate fifteen-minute time limit, were unable to undertake roles of help beyond the confines of the family. This latter group presumably includes many of the "dazed" persons described by early observers. At impact-plus-5 minutes, probably a larger percentage of uninjured persons would appear to be dazed; at impact-plus-30 minutes, no doubt the percentage would be lower. In other words, the percentage of "on-helpers" is a function not only of susceptibility to a "dazed" reaction but also of the time elapsed after impact (and, looking beyond the Worcester event, perhaps also of such situational factors as the suddenness, unexpectedness, duration, and destructiveness of impact).

It is difficult to comment on the patterns of leadership in the impact area during the isolation period at Worcester. The data given above suggest that during the first fifteen minutes after impact very few persons assumed roles of leadership responsibility (i.e., community-oriented roles) who had not held roles of formal responsibility to the community before impact. This seems on the surface to be at variance with the frequently reported post-disaster phenomenon of "spontaneous" or "emergent" leadership by previously inconspicuous persons, partly to fill roles functionally emptied by the injury, absence, or failure to perform adequately of many "official" leaders.

The Worcester situation, however, made emergent leadership functionally almost unnecessary, since (as will be developed later) the isolation period was short and the community counter-measure agencies were intact. Furthermore, during the first fifteen minutes after impact, persons who had leadership roles already assigned to them probably found it easier to act as leaders than potential "emergent" leaders, who probably require a longer period of orientation because they need to orient themselves not only to a new situation but to a new role.

In other words, even in situations, unlike Worcester, where spontaneous leaders do ultimately become very important, the first leaders to act may be those who already occupy leadership status.

MAP 7: Disaster Space at 5:30 P.M.

- ･･･ total impact
- ─── fringe impact
- ─·─· filter area
- ─ ─ ─ organized community aid (as yet, notification and mobilization of community are fragmentary)

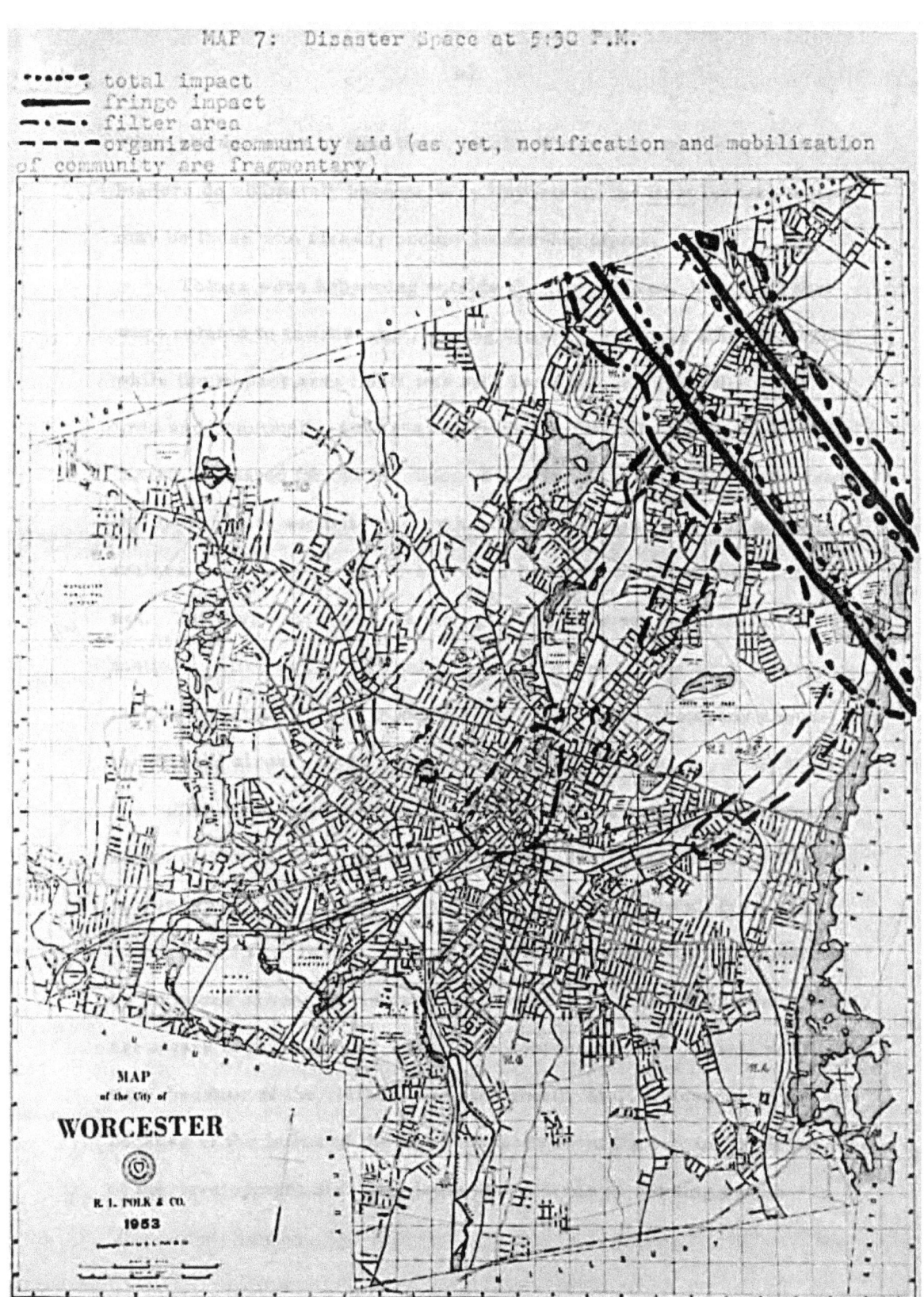

Things were happening outside the impact area, however, that were related to the disaster, during the approximately fifteen minutes while the impact area itself was still isolated. First of all, the filter area and community-aid areas were differentiating and assuming their disaster-related functions. Secondly, within the community-aid area, the appropriate security agency headquarters, stations, and administrators were being notified, and were in turn mobilizing their personnel. Thirdly, various operating agencies were setting themselves in motion: police cruisers, ambulances, and fire trucks were already on their way, or even (as for example, with the ladder company above-mentioned) already were approaching or at the scene.

The differentiation and role-assumption of a filter area depended not on official notification or mobilization, but on the virtually simultaneous influx into the area of visual and auditory stimuli from two directions. The filter area may, in a negative sort of way, be defined as that zone around the impact area, virtually all the residents of which are aware of the disaster within five minutes, without formal notification, because of the visible, audible, smell, etc., evidences of impact; because of the influx of the first evacuees from the impact area; because of the development of traffic jams as hundreds of working people – in many cases, the parents, children, or other relatives of impact area people – in automobiles piled up at the barricades of trees, structural debris, and live wires at the edges of the impact area; and because of the influx of security vehicles and apparatus (police cruiser cars, patrol wagons, ambulances, and fire apparatus), much of it equipped with sirens and blinkers, also piling up in the traffic jams until both private cars and debris had been cleared away. Thus the filter area was saturated, as it were, with disaster stimuli, from both directions; through it passed communication, equipment, and personnel, at maxi- mum concentration. For the filter area, the stimulus was multiple, continuous, and preoccupying; in other parts of the city, it was selective, discontinuous, and "news."

During the isolation period, to judge from impressionistic accounts, filter-area people played a variety of roles. They were in a peculiar situation of knowing that something serious had occurred, without quite knowing what it was, and a number of filter-area people walked into the impact area to find out. These people were immediately tagged as "sight-seers." During the first fifteen minutes, however, probably most of the so-called "sight-seers" were bona fide residents of the impact area returning from work. Some filter-area people were beginning to step out of their homes to serve as volunteer traffic police. Some were giving asylum or transportation to injured refugees. Some were running into the impact area to engage in rescue work. Some were giving first aid. Some were "rubber-necking" and getting in the way. Some were looking out of the window and wondering what was going on. And some were cleaning up around their own homes after the severe hail and thunderstorm which had struck them. A few had seen the tornado strike; others had not.

The fundamental characteristic of the filter area is that it is that part of the non-impact region, adjacent to the impact area, where notification of the disaster having struck is quick and disseminated to almost everyone, which can provide the initial manpower for rescue and evacuation, and through which much communication and transportation must pass.

Beyond the filter area lies the "community aid" area. In contrast to the filter area, notification to the community aid area is selective: the specific security agencies (police, fire, hospitals, civil defense, and Red Cross) are notified at once; they notify and mobilize their personnel; and the general populace is not "notified" (except by beginning ripples of word-of-mouth and telephone talk) until radio or other communication media begin to broadcast appeals for blood, or accounts of the disaster, about an hour after impact. The community-aid area also gets its notification, not by contact-perception, but through signals and symbols.

The time of notification of Worcester security agencies is difficult to establish. Two sources of notification were built into the standard operating procedure of the agencies themselves: several police cruiser cars, with two-way radio, had "beats" crossing or intersecting the impact area; and fire alarm boxes were located in the impact and filter areas. Although the data have not been tracked down, some police cruisers no doubt radioed information to police headquarters (and anyone else listening). This is probably the source of the information received by ladder company #10, who were monitoring a radio and picked up a statement that a boiler had exploded at the Home Farm, and were on their way while a box alarm from the same area was still coming in. In my opinion, a combination of box alarms and cruiser calls probably had alerted both fire and police headquarters within five minutes after <u>initial</u> impact in north-western Worcester; by five minutes after terminal impact in eastern Worcester, these headquarters probably had a picture of the general location and course, but probably not width, of the impact area.

But box alarms and cruiser calls were not the only sources of notification. Telephones were still working between at least the southern fringe area and the center of Worcester, and some calls were probably coming in. Furthermore, the State Police at Holden (which was struck about 5:00 PM, at least fifteen minutes before terminal impact time in Worcester) had gotten on the radio at once and informed State Police Headquarters in Boston. Massachusetts Civil Defense, with its Headquarters in the same building with the State Police, called Region 3 (containing Worcester) Headquarters, who notified the regional director's administrative assistant, sometime during the impact at Worcester (but of course without mentioning Worcester as an impact area). Worcester Civil Defense Headquarters was empty (the city director was caught in a traffic jam in the filter area, and the secretary had gone to her home.) By 5:30, however, the regional administrative assistant was at Worcester CD Headquarters, attempting to mobilize local and regional CD units, and to find out what had happened.

The hospitals, likewise, were gaining their first knowledge of the disaster during the isolation period. The telephone call to City Hospital at about 5:20 has been mentioned. About 5:25 a man with a lacerated scalp walked into the courtyard of Hahnemann Hospital. By 5:35 ambulances were arriving at Memorial Hospital. And, of course, the wail of sirens from police, fire, and ambulance s (under police control) must have alerted the hospitals (as well as much of the rest of the city) to the fact that something unusual and disastrous had happened.

Thus, during the isolation period of perhaps an average of fifteen minutes' duration, notification, inventory, and mobilization procedures were under way in three functional areas: the filter area (notification being almost total through the population, by sight and sound); the community (notification of security units, particularly fire and police departments, by several channels — telephone, box fire alarms, radio from radio cruisers, and siren sounds); and "outside" security units (State Police and State and Regional Civil Defense, in particular being notified by radio and telephone).

Police and fire were, during the isolation period, mobilized, dispatched, and penetrating parts of the impact area. Notification of hospitals and Civil Defense, however, was slower, it "took" less well, and CD in particular was, through no negligence of personnel, almost if not completely inactive during the isolation period, with its director caught in a traffic jam in the filter area, and its office closed for the day; the office was actually opened by the assistant to the regional director, who got his word about a tornado in Holden (not Worcester) via a chain of radio and telephone messages through Boston, and to whose intuitive reaction is owing the fact that the Worcester CD office doors opened before 5:45 (when the female secretary-and-telephone-operator came in).

V. RESCUE: Extrication, First Aid, Reassurance, And Evacuation

With the arrival of police and fire department personnel and apparatus, the isolation period came to an end, in most parts of the impact area, by 5:10. What happened after that is considered to constitute a fifth period of disaster time: the rescue period, lasting through the night of June 9-10.

Rescue and evacuation operations were conducted mainly by City of Worcester police, fire department, and public works personnel, including both regular employees (off-shift as well as on-shift), auxiliary firemen and policemen (many of whom were enrolled as Civil Defense auxiliaries and wore CD armbands, but had been called out by the heads of the Police and Fire Department Auxiliaries, respectively). These people worked under loose police department supervision, more or less methodically going up and down streets freeing trapped persons, giving minimal first aid and encouragement, and identifying places where rescue equipment (bulldozers, winches, cranes, etc., would be needed). These people constituted a nucleus of relatively well-trained and purposeful people about whom collected various random volunteer assistants: residents of the impact and filter areas, teenagers, "sightseers," et al.

Meanwhile, all available ambulances were assembling at various points in the filter area and in the fringe impact area. Most of these ambulances were under the control of the police department; Worcester hospitals did not control the ambulances. But at least a few non-police-controlled ambulances also showed up, volunteered by funeral directors, industrial plants, etc. These ambulances evacuated such people as were carried or could walk to them; rarely did they penetrate deep into the impact area. Private cars, trucks, and station wagons, operated by impact or filter area residents, drove many victims from their site of injury to the ambulances. Many persons were taken directly to the hospitals from the impact area in trucks, cars, and station wagons.

The traffic jam in the filter area was, by 5:30, formidable, and was interfering with the passage of fire, police, and public works vehicles, and ambulances. Regular police were occupied with rescue and evacuation procedures, but according to several reports, "spontaneous volunteers" took over the job of directing traffic in and around the impact area. These "spontaneous volunteers" may actually have been auxiliary CD police. Complicating traffic problems, of course, were the hundreds of fathers, mothers, sons, and daughters of residents of the impact area, abandoning their cars in the filter area and running into the impact area on foot to find and help their families. Mass communication devices were brought into use by police: loudspeakers mounted on police vehicles were employed, for example, to plead with residents not to smoke, in areas where gas mains were broken. Although most radios in the impact area were out of commission, a few persons heard (by radio or by rumor) the actual broadcast at 7:00 PM of a tornado warning.[24]

---

[24] Interview with Mr. Brisette, of WTAG. The use of loudspeakers to quash or explain rumors is not reported.

There are conflicting and confusing reports about the extent and quality of first aid performed in the impact area. Estimates by hospital personnel are that less than 10 per cent of casualties brought to the hospital had received first aid. The evacuation procedures made classification, routing, and expediting of casualties almost impossible, even if there had been medical personnel to perform triage functions in the impact and filter areas. It would seem that actually there was a great deal of first aid of a primitive sort administered by victims and their relatives and neighbors, by police and fire personnel (who were trained and whose vehicles carried first aid kits), and by a few physicians and nurses who were working (without much in the way of supplies) in the impact area.

Wounds were bandaged, some tourniquets applied, a few fractures splinted, a few persons given unidentified "injections" (sedatives to mitigate shock?). It is of course almost impossible to estimate the effect of what might be called the negative canons of first aid, involving not doing the wrong thing even though little therapeutic aid was given.

Various formal first aid stations, with equipment and supplies, doctors, and nurses were set up: one at a drug store on Burncoat Street in or near the filter area; another was set up by a hospital in the backyard of one of the homes; another was set up at Curtis Apartments; and still another at Municipal Auditorium. Generally speaking, however, major injuries did not go to the first aid stations but were taken directly to the hospitals; indeed, most of the seriously injure ed were evacuated in two or two and a-half-hours – just as the first aid stations were getting organized. These aid stations tended, therefore, to care only for minor injuries.

The fire and police departments in the first hour or two of the rescue period had a major responsibility (particularly the fire department) in combating secondary impact. There was material secondary impact of four major kinds: live wires all over the place; actual fires in buildings; escaping (and inflammable) cooking and heating gas; glass and other hazards in the way of pedestrians.

The power company substation on Boylston Street turned off the power to much of the impact area within a minute after impact in that area, but this was not general knowledge, so the presumed live wire problem had to be met by stationing guards, helping or guiding people around or over the wires, and in some cases chopping sections apart with rubber-handled axes (although this was risky where sparks might ignite gas)[25].

---

[25] Powell, interview with Provost, p. 8.

MAP 8: Disaster Space at 6:00 P.M.

```
..... total impact
────  fringe impact
─·─·─ filter area
──── community aid (mobilization virtually complete)
ooooo community aid (mobilization still fragmentary)
```

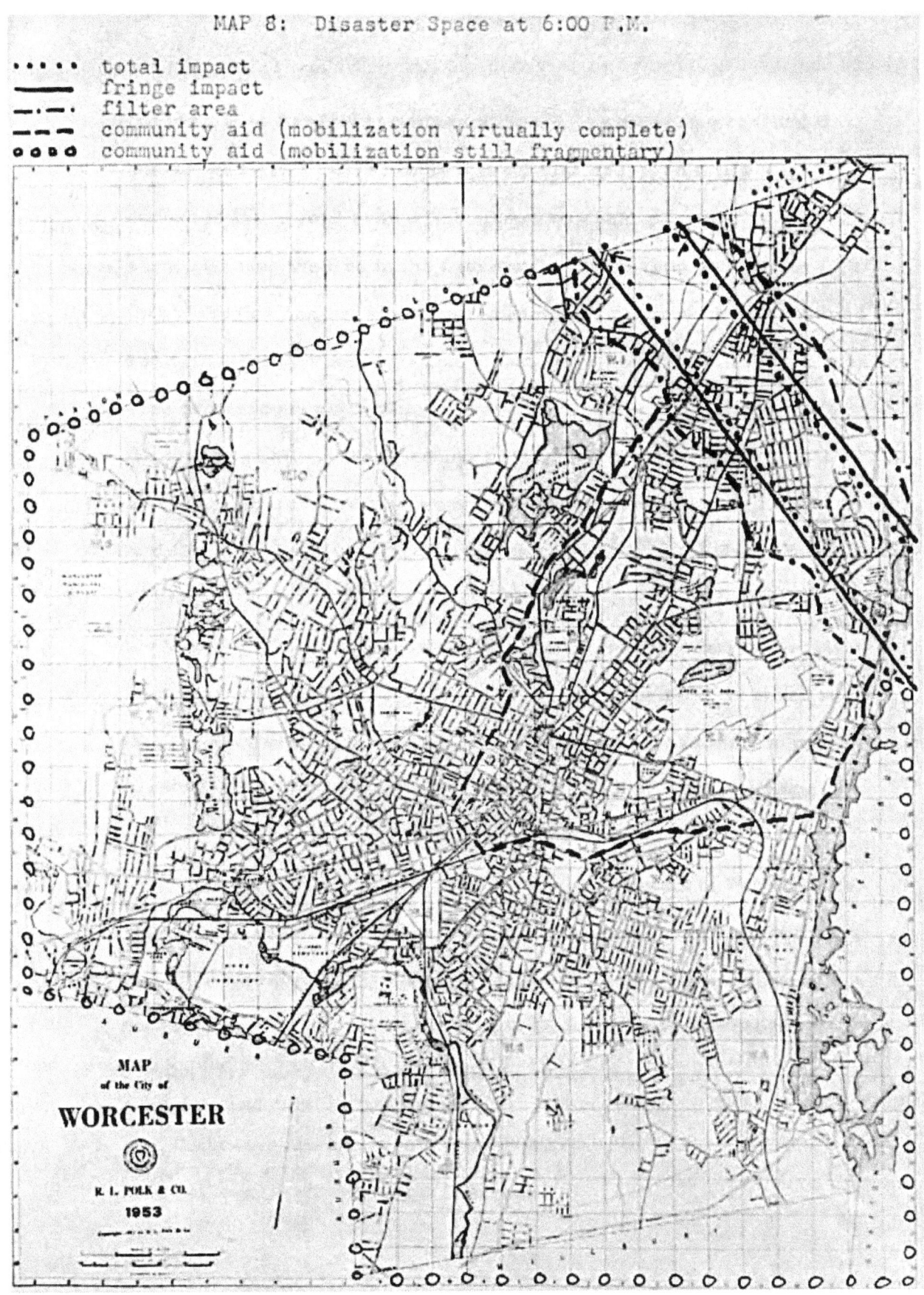

No deaths or injuries are reported to have occurred as a result of contact with live wires during the rescue period. At least nine buildings burned completely or in part; the total number was probably more. An industrial building on Brook Street, in the Industrial Area, was gutted, although fire apparatus reached it during the fire[26]. On Francis Street, four houses in a row were burned to the ground before fire apparatus could get water to the scene and stop the fire in the fifth house.

Some reports indicate that if the fire had not been stopped at that point, it would have spread and turned into a "conflagration," because it would have moved into an area of wooden housing where there was no water in the hydrants (at the top of the hill in the Burncoat area)[27] A building in the Brattle Street area burned, although apparatus got there during the fire[28]. And there were at least three fires which started in the Great Brook Valley section but were put out quickly by the fire department.

While the story of these (and probably other, not-reported) fires is not given much space in the various narratives which I have seen, it would seem that the quickness with which fires were either extinguished or prevented from spreading was an extremely important factor in reducing casualties and damage below their potential level.

A third Worcester city agency – the Department of Public Works – played an important if somewhat imponderable role.

As has been indicated already, the traffic blockage at the southern boundary of the impact area was the result of both an accumulation of debris, particularly boards, wires, and fallen trees, and of passenger vehicles.

Fire equipment literally hacked its way through and over these roadblocks in some places, but ambulances and lighter vehicles generally could not follow. Although I do not have much data on this, it would seem that the Department of Public Works must have quickly dispatched crews to the area with bulldozers, chain saws, and other heavy equipment, to clear the roads.

This was not, however, reportedly completed at many points Until an hour or an hour and a half after impact, and then only one-way passages were available.

This meant that the impact-area edge of the filter area was a filter in a physical sense, tending to block movement in either direction.

Thus within the period from impact to 8:00 PM, all severely injured persons (but not all bodies) had been removed from the impact area; secondary impact from fire, gas, and live wires had been controlled; one-way traffic lanes had been cleared through the debris and traffic jam in the filter area; and CD auxiliaries, city, and state police were on duty at the perimeter, manning traffic blocks and controlling traffic. Activity up to 7:00-

---

[26] Callaway, Interview with George Murphy, p. 10
[27] Powell, interview with Johnson, p. 11.
[28] Callaway, interview with Sharrocks.

7:30 was, apparently, chiefly Worcester activity: that is to say, such resources from other communities as were in the process of mobilization had not yet arrived.

Thus, for example, one hour and twenty minutes after impact (i.e., between 6:30 and 7:00) the only two rescue trucks under the control of Massachusetts State Civil Defense arrived at the impact area. It is difficult to estimate how important these two rescue trucks were in the total schedule of rescue work: apparently they carried the only proper rescue tools which reached the impact area (Worcester CD reportedly left its rescue tools in a warehouse, having no means of transporting them to the area and no rescue workers trained in their use). The trucks were, when they arrived, manned by three trained men (two drivers and the Chief of Rescue for Massachusetts); Worcester's Department of Public Works supplied fifty men. These two trucks and their crews and equipment worked the next thirty hours until all known casualties were released. It is not easy to estimate, from the report, how many live casualties they released; my impression, however, is that almost all live casualties had already been released, and the Rescue Service was chiefly occupied finding the bodies of missing persons[29]

Thus the period between about 5:30 and 7:30 constitutes a clearly defined time during which standard community protective agencies (police, fire, and public works departments personnel) were, with the assistance of a large number of volunteers who entered through the filter area, and with the aid of victims themselves (who seem to have been stimulated to greater activity by the arrival of uniformed outside aid), able to:

- neutralize secondary impact (fire, gas leaks, live wires);
- give extensive, however crude, first aid;[29a]
- rescue from entrapment all living trapped victims;
- evacuate.to hospitals all seriously injured persons.

Such coordination as these activities achieved was exercised by the police, but it is evident, from the interviews with victims, police, and fire department personnel, that very little planning and organizational work was done. The agencies operating were all highly trained teams who generally followed a standard operating procedure, modified as circumstances required; like a military unit comprised of veteran troops who have had experience working together under fire, they could function without extensive direction.

The untrained thousands of victims, absent residents who returned to the area after impact, and volunteers from the filter area were also able, by sheer mass of manpower, to accomplish a great deal, even if per capita efficiency was low.

Thus a mass of untrained volunteers helped a fire crew to lay a long hose faster than such a hose had ever (in their experience) been laid before. Such crews of rescue workers were able to pull out the less seriously trapped at once, leaving the severe jobs for the trained teams with equipment.**xx**

---

[29] Knight, 1953.
[29a] in the estimation of trained medical personnel, most of this "first aid" was of negligible value and not worthy of the name.

Plate 5: RESCUE AND FIRST AID IN GREAT BROOK VALLEY

Plate 6: EVACUATION FROM GREAT BROOK VALLEY

Dozens of vehicles carried casualties to hospitals, roaring down Burncoat and Main Streets at sixty miles per hour with horns blowing.

While that sort of mass rescue and evacuation, with professional aid at critical points, no doubt was inefficient in many respects, it was perhaps as effective on the whole as a better organized but slower operation. The Worcester rescue and evacuation can be looked on as a kind of specimen of optimum good fortune in the face of disaster.[29b]

A major aspect of the rescue period (although it continued, in the hospitals, until the morning of June 11) was the emergency hospital medical care of injured persons. First aid in the impact area has already been mentioned. Conceptually, this should be distinguished from the long-term hospital care of seriously injured persons. The problem of emergency care is to prevent death or intensification of injury, and involves such procedures as triage, stopping of bleeding, combating of shock, transfusion of blood or plasma, administration of tetanus antitoxin, setting of broken bones, cleaning of wounds, etc.

Subsequent hospital procedures should, from a sociological point of view (if not necessarily from a medical viewpoint), be regarded as part of the rehabilitation process.

The hospitals in Worcester, as was indicated earlier, do not own and dispatch the ambulances, which are under the police department. Furthermore, the hospitals – although, of course, theoretically related to Civil Defense – are not explicitly part of the civil defense organization. Some of them (notably the City Hospital) had disaster plans, but these were generally in a paper rather than stand-by state of organization, and the plan (if any) of one hospital was not related to that of other hospitals. This placed the hospitals as a group out of the main line of communications among the protective agencies: CD, police, fire, and Red Cross.

City Hospital – the largest general hospital in Worcester – had a disaster plan; it, furthermore, received a telephone call or calls about 5:15 or 5:20 (probably the call from the housewife in Great Brook Valley, and a call from the City Police to the effect that lightning had struck the Home Farm) which indicated that a number of casualties were going to arrive. The Superintendent called the police to check on this advice.

But at 5:40, before any disaster plan could be set in operation, City Hospital received its first casualties, and within minutes it was deluged. A number of the non-resident medical staff came to the hospital on their own initiative within about half an hour. Patients arrived in ambulances, cars, and trucks. Newspapermen swarmed into the building and pre-empted the telephone (as they were to do at Civil Defense Headquarters). Two hundred and fifty blood donors crowded into the blood bank. There were not enough police – not even auxiliary police – to direct jammed traffic; one policeman on duty there that night died of a coronary thrombosis the next day. Families of the injured descended on the hospital in droves, and the hospital – in spite of considerable resentment – was forced to exclude relatives.

29b. Observers often are surprised, apparently, by this "luck" component in the rescue period.

MAP 9: Disaster Space at 8:00 P.M.

· · · · · total impact
━━━━━ fringe impact
━·━·━ filter area
━ ━ ━ community aid area (notification and mobilization nearly complete)

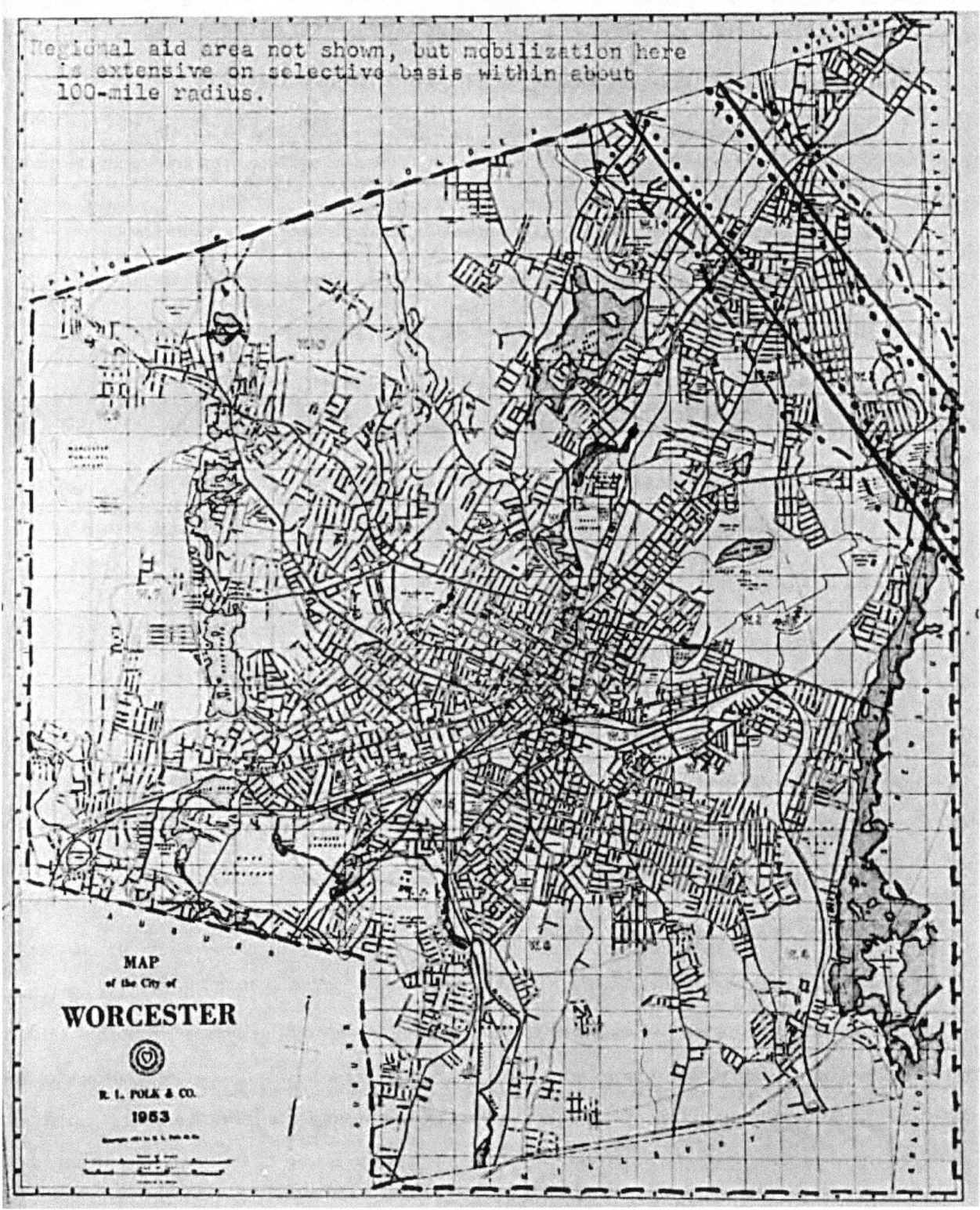

Regional aid area not shown, but mobilization here is extensive on selective basis within about 100-mile radius.

All in all, in the words of the superintendent, during the first half hour there was chaos. This chaos gradually straightened out during the night. Patients who could be evacuated were sent home to make space for tornado victims: only eighty could be admitted the first night, and about forty more the next day. In addition to these given beds, about two hundred and fifty more were treated at the hospital that night. There were insufficient supplies of blood fluids and of tetanus antitoxin: Boston and New York were telephoned for these. Volunteer doctors and nurses had to be assigned; triage systems set up; a. headquarters set up.

After six hours, things were sufficiently under control for anxious relatives to enter the hospital. No communications of a dependable sort were established even by the next day, however, with agencies in the disaster area.[xxi]

Plate 7: EVACUATION FROM BURNCOAT STREET AREA

Similar problems were met at other hospitals. At Hahnemann, close to the impact area, the first indication of a disaster was the sight of a man with a lacerated scalp, his face covered with blood, who walked into the yard about 5:15, presumably from nearby Burncoat area. The director of the hospital, who was at dinner, sent an X-ray technician down to him. When she looked out of the dining room window again, "the yard was full of people streaming in all the doors." Thereafter it was the same story as it was at City Hospital: streams of patients, traffic jams in the driveways, staff assembling spontaneously, telephone calls for tetanus anti-toxin, blood, and plasma, masses of volunteer workers and blood donors, newspaper reporters underfoot.

And similar accounts were given by representatives at other Worcester hospitals, stressing the same general sequence of events: a mass of injured people suddenly piling up at the doors, without warning (at Memorial, which received more patients than any other, the only warning was the sound of sirens on the ambulances and cruisers which arrived about 5:35); spontaneous mobilization of doctors, nurses, and volunteers; calling in supplies as stock piles were exhausted; lack of communication with other protective agencies.

The general impression of the activity of the hospitals during the emergency night is of a group of isolated institutions, unprepared as to plan and totally unwarned, suddenly deluged with patients, and thereafter improvising organization as they went along. Because there was an over-abundance of doctors, nurses, and volunteer workers; because the community as a whole was intact; and because supplies could be obtained readily from nearby communities, administrative difficulties which might have led to deaths or complications of injuries or a number of patients were overcome by the sheer mass of personnel and material available.

The load of patients was distributed among the dozen Worcester hospitals as indicated in the breakdown of hospitalizations:

TABLE 7: THE BREAKDOWN OF HOSPITALIZATIONS	
Memorial Hospital	168
City Hospital	103
Hahnemann Hospital	55
St. Vincent Hospital	27
Fairlawn Hospital	10
County Tuberculosis Sanatorium	8
Doctors Hospital	3
Harvard Private Hospital	0
Emergency Hospital	0
Belmont Hospital	0
State Mental Hospital	0

Thus, of the 374 casualties requiring hospitalization, 326 or 87 per cent were cared for at three of the eleven hospitals. It would seem reasonable to suppose that, even though the other eight hospitals may not have been prepared to cope with a large number of casualties, they could have given more seriously injured victims care, and thereby have reduced the burden on the three hospitals which received the major deluge of patients.

An extensive report on medical care in the Worcester tornado has already been prepared, and there is no reason here to recite the details there presented.[xxii]

That report, however, indicated that the emergency medical care was deficient at two points: in the handling of wound shock, and in the handling of open wounds.

In the former case, there was reportedly a low incidence of wound shock observable in victims in the impact area, but a high incidence at the time of arrival at the hospital. The reasons for this are not developed: one might suspect that the helter-skelter evacuation, involving the handling of many patients by inexperienced persons, their hasty placement (in many cases) in crowded trucks, station wagons, and private cars, and their transit to hospitals at extremely high speed, may have in some cases intensified or accelerated the development of incipient physiological shock symptoms by "infecting" the patients with a sense of anxiety and lack of control on the part of the rescuing community. Furthermore, for some reason very little whole blood, plasma, or plasma extenders was administered, although this would have been an appropriate anti-shock treatment.

In regard to the handling of open wounds, many lesions were sutured immediately after cleaning. A lower incidence of infected wound sites would certainly have been achieved if delayed closure had been the method followed.

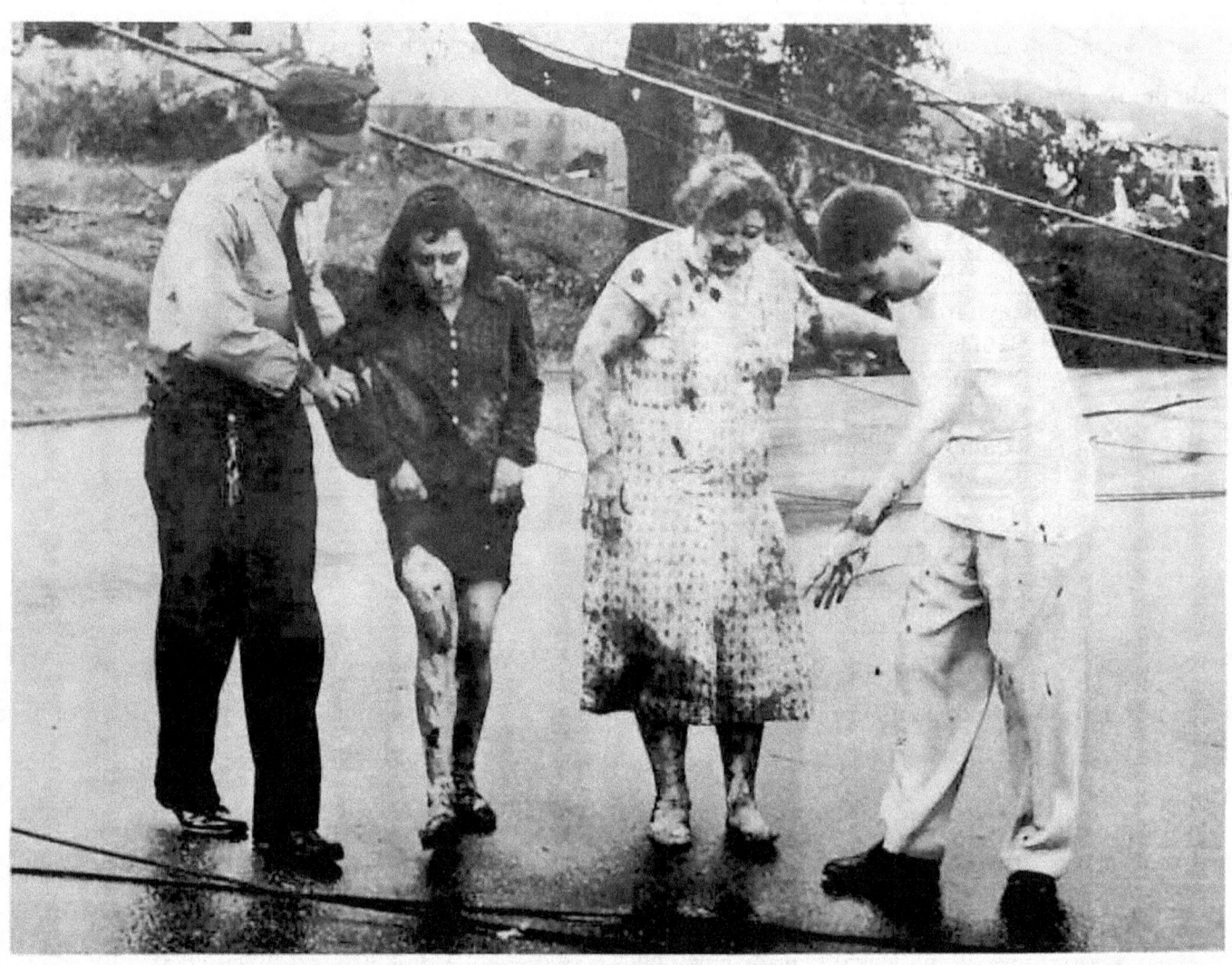

Plate 8: EVACUATION FROM GREAT BROOK VALLEY

VI. REHABILITATION: The Attempt to Restore the Steady State

Each succeeding stage of the disaster involved a wider area. The impact period involved only the impact area; the isolation period involved the impact area, a filter area, and selected community institutions, particularly the fire and police departments; the rescue, evacuation, and emergency medical care period involved all these, plus a wider panel of institutions within the community, a large number of volunteer workers and blood donors in the community, and institutions outside the community selected for their organizational relationship to Worcester protective units: regional and state CD, state police, other hospitals and medical supply houses, etc.

The rehabilitation period is the longest, the widest in area of involvement, and the most complex in interpersonal relationship. It is certainly the one which obtains the most massive publicity. In the rehabilitation period, the disaster ls a given, and there is a tendency for attention to focus on the leaders of rehabilitation agencies as the heroes (or villains) of the whole disaster; the role of victims, and of the protective agencies who arrive first (police and fire departments) tend to be ignored. Furthermore, rehabilitation agencies involved in a disaster of the magnitude of Worcester's tend to be concerned (of necessity) with public relations, and to blanket the media of mass communication; and still further, they are well enough financed to be able to make extensive records, surveys, self-surveys, etc.

In the case of Worcester, all of this means that while usable data on the isolation period, for instance, is sparse, there are great quantities of data on the rehabilitation period.

It will not be possible to do more than outline some of these data in this report; and on many areas of importance, I shall be unable to do more than to mention the area and make some brief indication of its significance.

Rehabilitation aims essentially at recreating the conditions that were; it assumes a prior impact, and a prior emergency period during which secondary impact has been stopped, but at the conclusion of which there is a new situation.

Rehabilitation aims at restoring personnel, in so far as possible (and desirable) to their pre-impact physical and emotional status; and at repairing and rebuilding the damaged material culture to its pre-impact status (again, as far as possible and desirable).

Complete recreation of the past is of course impossible: not all injuries are reversible, not all demolished structures can be replaced; and for various reasons, it may be decided not to recreate but to replace with something significantly different.

The rehabilitation procedures are directed at two separate but closely related ends: the restoration of individual health (emotional as well as physical), employment, and standard of living; and the maintenance and restoration of the material culture (furniture, dwelling units, public utilities, public buildings, roads, industrial and commercial structures, plant cover, etc.). Various instrumentalities come to be of crucial importance here, and may even become temporary ends in themselves: financial resources; the organization of participating institutions and individuals; the housing and feeding of rehabilitation workers; transportation and communication; and so forth.

The institutions which played particularly significant and dramatic roles in achieving these ends, after the Worcester tornado, were: the American Red Cross (local, regional and national organizations); Civil Defense (local, regional, and national organizations); the National Guard (181st Regiment); the Salvation Army; various industrial concerns, particularly the Norton Company; church organizations (both locally and on a wider level); Worcester city welfare, public utility, housing, police, and other agencies; insurance companies; innumerable suppliers, contractors, and their labor force; the hospitals and medical personnel generally; communications agencies (newspapers, radio, telephone, and telegraph); and various state and federal agencies concerned primarily with legal and financial organization and aid. The complexity of the interrelationships of this multitude of institutions was. far beyond the rational understanding of the participants themselves, and this report cannot begin to describe empirically and in detail the development and periodic conditions of these relationships, nor account in detail for all their activities, whether coordinated or not. Some of these institutions are still actively involved in rehabilitation work, a year after impact, and will no doubt continue to be, to a greater or less degree, for many years to come.

Let us first consider the restoration of the health, employment, and standard of living of victims from the impact area.

The rehabilitation period commences when the seriously injured have been evacuated and hospitalized; minor injuries have been given first aid at hospitals, by private physicians, by relations and friends, or by victims themselves; and the area of total impact has been almost completely evacuated.

Rehabilitation involves two complementary activities: prevention of any further impact, or deterioration as a result of processes initiated by primary or secondary impact; and the restoration of health, employment, and standard of living.

1. Hospitalization

In regard to health, several organizations and a multitude of individual agents become involved.

Most obvious, in the Worcester experience, were the hospitals, caring for the major injuries[30]. Of the approximately 804 patients who entered the hospitals for aid during the rescue period (some of these, however, probably came in a day or two after impact), only 374 remained as "permanent" cases for rehabilitation. The rest received any required follow-up care in their homes or temporary quarters. There was, however, an additional undeterminable number of persons who were major or minor casualties who never did reach a hospital;

---

[30] Information on hospital care during the Rehabilitation period comes from three sources: Bakst, et al, 1953; American National Red Cross, 1952; and interviews with hospital administrators by Miss Jeannette Rayner

these must be added to the 804 on hospital lists, and probably would bring the total number of casualties in Worcester alone well over the 1000 mark. My guess would be that the total number of persons requiring some sort of medical attention or first aid for major or minor injuries would, in Worcester alone, be 1600 - i.e., for every injured person who received hospital care, another was cared for extra-mural.

Hospitalization statistics certainly cannot be used as an adequate count of casualties[31]. It might be desirable in disaster situations for some uniform procedure to be set up to make possible an adequate count of casualties treated by physicians and dentists privately, by trained nurses, police and fire department personnel, etc., since these patients must play a considerable, if unrecorded, role in the logistical process.

I do not have statistics on the number of hospitalized Worcester casualties who remained in Worcester hospitals at successive dates after impact. One would expect, however, that the number of hospitalized casualties plotted by time would form a descending logarithmic curve, with the bulk of cases being discharged within a week or two, but with a few cases remaining for months.

In addition to the Worcester hospital staffs, private physicians, trained nurses, dentists, druggists, etc., who dispensed most of the care directly, several auxiliary organizations played important roles. The Worcester Society for District Nursing mobilized at least twenty nurses the first night of the disaster, and thereafter its nurses followed many patients in their homes or temporary quarters. Civil Defense operated an aid station in the Municipal Auditorium.

The American Red Cross mobilized the bulk of the necessary extra nursing care: 581 volunteer or Red-Cross-affiliated nurses worked under Red Cross direction (142 of whom were paid by the Red Cross) for a total of 3,002 days, up to October, 1953, caring for critically ill hospitalized patients. These nurses came from no less than thirty-nine Red Cross chapters!

The cost of this massive application of medical and nursing care to about sixteen hundred casualties in Worcester was undoubtedly very large, both in terms of personal services and facilities, and medical supplies used. But there was almost a competition over who was to pick up the check!

Within a day after impact, some of the local hospitals announced that they would send no hospital bill to any tornado victim. Then the Governor of Massachusetts was reported to have announced that the Commonwealth would pay all hospital bills; this report, however, was denied. The Red Cross then assumed responsibility for paying all tornado hospital bills on a patient-per-diem basis. All in all, the Red Cross paid for medical and nursing care for 432 families (this, however, includes other towns besides Worcester); probably in the end the Red Cross paid for the hospital and nursing care of all, or almost all, seriously injured victims.

---

[31] The statistics of casualties in Bakst, et al, 1953, are based only on the count of those who were recorded by the American Red Cross as having received some form of hospital care, although the authors indicate that not all casualties were treated at hospitals. No doubt all, or almost all, major injuries were treated at hospitals, however.

Up to the end of 1953, however, it would appear that the ultimate administrative source of funds had not been determined.[32]

In any case, the principle was evidently accepted that the burden of medical expense on the injured victim (who in most cases, if adult, had also suffered from loss of income, possibly of employment, and property) should be assumed by the community in one way or another, since individual responsibility would in almost all cases interfere with total rehabilitation of the impact area and would in many cases have been impossible to support. In general summary, it can be said that medical rehabilitation was largely defined as a community responsibility, and that rehabilitation procedures for those who were hospitalized were administered on the basis of medical need, not on the basis of individual ability to pay. What the practices of privately consulted physicians were is not indicated by my data.[32a]

But the general definition of the medical situation was that maximal rehabilitation was a community aim and responsibility; and that the cost of it should be borne by the Worcester community, the state and to some extent the whole nation.

2. **Welfare Services**

The impact area was very largely evacuated by 11:00 PM on the night of the tornado. This meant that there were about seven or eight thousand homeless persons in Worcester that night, to be given shelter, food, and in some cases clothing (as well as medical care for some of the injured). Some of these persons moved back into the impact area in a day or two: some (particularly those from the public housing projects and from completely destroyed homes) did not move back for several months; and some (comparatively few) never returned, having moved permanently elsewhere.

A multitude of organizations and persons became involved in the administration of the aforementioned welfare services. Notable among these were: The Central Massachusetts Disaster Relief Committee, appointed ad hoc to collect voluntary contributions; local, state, and federal Civil Defense: The American Red Cross; the Salvation Army; Protestant and Catholic church organizations: municipal and state welfare organizations; summer camps sponsored by various groups, the YMCA and YWCA, and the Boys' Club; and various industrial concerns.

Many private persons performed what essentially were welfare services for relatives and friends; many businessmen in various ways extended aid in the form of outright gifts, price reductions, and expediting of shipments; many private individuals gave time, money, clothing, furniture, etc.

It is notable, first of all, that of the approximately two thousand Worcester families displaced, very few had to be housed in the emergency shelters. The Red Cross, who took over the administration of welfare, including shelters, from Civil Defense on June 10, had housed a total of fifty-seven families (257 people) in dormitories

---

[32] CF. Bakst, et al, 38-39, and American National Red Cross.
[32a] According to later information, private physicians made no charge for their services.

at Worcester Polytechnic and Clark University up to August 1. On the first night there were a great many more who received emergency shelter but there are no records to indicate how many. By the next day, almost all evacuees had moved in with relatives or friends, or rented new quarters.[33]

The Red Cross (local chapter) is said to have provided food for about six thousand persons (perhaps three thousand of whom were in Worcester) during the first night. Thereafter relatively few victims required mass feeding. Red Cross and Civil Defense mobile canteens and field kitchens functioned in the impact area for weeks, however, giving food – ranging from coffee-and-doughnuts to full meals – to rescue, reconstruction and repair workers, and local residents who had not evacuated. At the peak of operations, the Red Cross had twenty mobile and stationary canteens in operation, manned and equipped by chapters in four or five states.

Several organizations concerned themselves with supplying victims with clothing, since many persons had lost all clothes except what they were wearing during impact (and these often were extremely dirty). Large quantities of both new and used clothing were distributed by the Red Cross, most of it at the Municipal Auditorium: 200,000 pounds of used clothing sent in (unsolicited) to the Red Cross in Worcester from all over the country; $15,000 worth of new clothing and 3,500 pairs of new shoes (worth $17,500) were also given out by the Red Cross.

The Red Cross also provided free legal aid and advice; its social workers went over individual and family problems in detail; it operated a registry of dead, injured, and evacuated persons, and handled inquiries concerning these cases, with the particular aid of the Worcester HAM radio network; it ran a housing reference bureau: all services which, while transient and difficult to record, certainly played an extremely important role in the welfare part of the rehabilitation process.

Both Catholic and Protestant churches and church-supported organizations busied themselves in relief work, particularly in the matters of preparation, collection, and distribution of food and clothing. Church welfare policy had not been planned in advance, and there was often some uncertainty as to whether a given church or church organization was primarily responsible to its own congregation and faith, or to the community. Churches seem generally to have helped first "their own," then others.

A notable incident in the history of welfare activities was the struggle over trailers. The Federal Civil Defense Administration had authority to allocate 500 trailers which had been used in rehousing evacuees of the Kansas-Missouri flood. The Chairman of the Board of the Worcester Housing Authority (all three of whose projects had been hit, requiring evacuation of 891 families) announced that Worcester would request that the 500 trailers be brought to Worcester and installed at Lincolnwood. In spite of the low incidence of mass care applications (as described above), and the availability of several hundred vacant dwelling units in the city, the

---

[33] The function of kinship ties in shelter allocation in Worcester is particularly interesting in view of the commonly stated sociological assumption that extended family ties are relatively unimportant in urban life in America. The extended family (consisting of several related nuclear families) was extremely important in providing shelter for the victims of the tornado in Worcester.

city persisted in its insistence. The Federal Civil Defense Administration and Red Cross observers indicated that there was no pressing local need for these trailers. Nevertheless, after the announcement, the Red Cross received 665 applications for trailers (chief complaint about temporary quarters: nervous tension aroused by overcrowded living conditions), and 425 trailers were finally brought in, set up at Lincolnwood or on private sites, and rented or provided gratis.[xxiii]

This move led the Red Cross to invest in special kits of furnishings and equipment for trailer living, with which the trailer residents had to be supplied, at a cost of $35,000. It was the feeling of some observers that the pressure for trailers was generated chiefly by the desire of certain local officials to demonstrate dramatically, by bringing the trailers into town, that they were doing their best for the sufferers; on the other hand, the trailers were used, and their use may have relieved a difficult housing situation which the fewness of applications for mass shelter accommodations did not adequately reflect.

At the same time that medical and welfare activities were working toward, first the maintenance, and second, the partial rehabilitation of individuals affected by the tornado, extensive procedures were under way toward the physical reconstruction of the area, since only with this could any real rehabilitation, even on an individual level, be accomplished.

In the fringe impact area, where homeowners and tenants could repair minor damage, physical rehabilitation began, for some, almost at the cessation of primary impact, residents getting out with hammer and nails, brooms and pails, ladders, saws, and shovels to "clean up the place." Indeed, even in the area of total impact, there was a tendency for dazed persons to putter about, sweeping, mopping, and sorting out belongings from the wreckage, even though such activities were so trivial in comparison with the task to be done that their significance lay evidently in their symbolic meaning rather than their practical utility.

Estimates of physical damage by various agencies were under way the day after the tornado. It was reported that 50,000 insurance claims were anticipated by the various property damage companies covering policies in the impact area. Insurance, however, did not nearly cover the total amount of damage, and several other sources of aid were invoked: compromise settlements between victims and institutions holding mortgages on their homes; Reconstruction Finance Corporation loans; and direct grants from the American National Red Cross to cover the difference between the cost of repair or replacement, and what could be raised from insurance-payments and loans.

I do not have data on the role played by personal savings in paying for repair and replacement of property. Within a few days many crews of construction workers were swarming over the impact area (which was being policed by National Guardsmen, Civil Defense volunteers, and regular and auxiliary police), demolishing unsafe buildings, replacing roofs, windows, and walls on damaged-but-repairable structures, and building new structures from the ground up. Supplies of glass, roofing materials, lumber, nails, hardware, etc., and the labor to use them, were brought in from all over New England and adjacent parts of New York. The ultimate result, a year later, was an almost complete physical rehabilitation of the entire area: Norton Company and the rest of the industrial area completely rebuilt; Assumption College's main building almost completely rebuilt, with improvements; the two permanent public housing projects (Curtis Apartments and Great Brook Valley Homes) completely repaired; the churches rebuilt or repaired as required; and most of the private homes, stores, etc., either repaired or rebuilt. A few signs of the tornado remained: trailers on the site of Lincolnwood; bare foundations at the Home Farm; a few empty lots here and there, and a few battered houses, still unrepaired. An uninformed observer would probably not guess that a tornado had passed through the area, and might think only that there had evidently been a lot of new building, in the course of which some old structures had been demolished and not yet replaced.

The most important single observation about the rehabilitation period at Worcester, to this writer at least, is that there was a very large concentration of "therapeutic" agents at the site of the community's wound. There was, indeed, more of everything available than was ever brought into play. In terms of hospitals, there were more hospitals in the town than were ever used. The Red Cross itself mobilized some thirty physicians and nearly six hundred nurses; these were in addition to those already mobilized by the community. More clothing was sent in than was used. Even though only fifty-seven families actually asked for public assistance in housing (in mass shelters), 425 trailers were brought into the town.

The Red Cross alone supervised about 960 volunteers; it budgeted $1,020,000 for relief, and spent $1,016,162, assisting 1,290 families in one form or another, making 755 cash awards to families averaging $879 per family. Rehabilitation aid extended to the point of the Red Cross buying twenty-five new vacuum cleaners (at one-third cost), and loaning them to 300 families to clean their homes and furnishings – one cleaner being used, on the average, by only twelve families. Many of the organizational problems of the rehabilitation period were not so much problems of competition for scarce supply. but problems of selection or dovetailing of alternatives offered: trailers versus existing, if overcrowded dwelling units. State or Red Cross financial responsibility for hospital bills, arranging of respective roles of insurance, loans, and cash grants in rebuilding, etc.

One might say that the area was stacked high with people and materiel trying to be of service. This profusion of help should be emphasized for two reasons: because it be expected in most natural or accidental disasters in this country; but it cannot be counted on in the event of atomic or hydrogen disasters as would very likely occur in case of a major war. It therefore raises the serious question of whether the same kind of planning (including organization) can apply both to natural disaster preparedness and hydrogen bomb attack-preparedness.[xxiv]

VII. IRREVERSIBLE CHANGE: The City Achieves a New Equilibrium

A subject of considerable interest, but one on which very little data is available, is that of the irreversible changes which occurred in the culture of Worcester as a result of the tornado. As was indicated in the preceding section, the resources not only of Worcester itself, but of the state and nation were devoted to a mass rehabilitation program aimed essentially at restoring the equilibrium state which obtained before impact.

Nevertheless, some systematic changes undoubtedly did occur: some of them changes which would have been reversed if that had been possible; some of them changes which were deliberately made, either with a view to using this opportunity to make improvements which would have been desirable in any case, or with a view to making better preparation for any future disaster. Conceptually, "change" in this discussion means change in the system rather than in individual units. Thus, the replacement of a roof, a wall, or even a whole building with another which serves the same functions is not a change in this sense: the system is the same, although one unit has been substituted for another. Where a new building which performs a substantially different function replaces an old one, or where a site is left vacant, or a structure left unrepaired, then one can speak of a systematic change. The same principle, of course, applies to individuals: scar tissue over a healed wound does not represent a change, on this level of discourse, but any permanent functional disability, physical or emotional, does.[34]

It is necessary, for practical purposes again, to choose a terminal date for the rehabilitation period, and operationally to define changes remaining at that date as "irreversible". Let us take one-year post-impact date as the terminal date of the rehabilitation period, and therefore consider the period of irreversible change as beginning then.

The most obvious category of irreversible changes are those changes which have taken place in individuals as a direct result of the primary and secondary impact. Thus a certain number of persons have died – dropped out of the social structure entirely – and required a corresponding readjustment in many social units. Bereaved relatives and friends, and their mutual relationships have undoubtedly changed in consequence, for while an employer can in most cases replace a lost worker, a wife and children cannot replace a lost husband and father, in a functional sense. There can be no real reconstruction of the status quo ante here. The consequences of these deaths cannot of course be described properly without a slow and painstaking case by case analysis, and probably these individual cases would in their various features be similar to cases of non-disaster-associated deaths. Their significance in the category of irreversible change lies not so much in the unusualness of the

---

[34] It is apparent that the distinction made above is a rough and practical rather than a philosophically valid one. If one took absolute continuity of units as zero change, and argued that any substitution of units, however trivial, involved some functional difference, however small, then the distinction I have made becomes one of degree rather than of kind, and the point of differentiation is recognized as the threshold of awareness on the part of the observer of a functional difference which corresponds to a difference in form.

individual instance as in the fact that sixty-six persons died at once, making a slight but noticeable dent in the population pyramid for the town.

Similarly, obvious cases of irreversible change are permanent injuries and disabilities, of various degrees of seriousness. The number of these probably exceeded one hundred. Here again individual cases no doubt presented the same features which might have been expected in the same number of cases of injuries resulting from industrial accidents, falls, etc.; it is the mass occurrence which represents the significant change in the system, since under ordinary circumstances they would have been distributed over a considerable span of time.

Significant changes in the economic circumstances of affected families and institutions were no doubt present, in spite of the efforts of such rehabilitation agencies as insurance companies, the Red Cross, state and local welfare departments, etc. I have no data whatsoever on these matters, and can only draw attention to the existence of the problem.

As indicated in the section on Rehabilitation, physical damage was largely repaired by April of 1954. Several dozen house sites where homes had been destroyed remained vacant, however, and a few homes had not been repaired. Assumption College's main building had been repaired, using what was left of the old structure, but the frame convent had not been replaced.[xxv]

St. Michael's on the Heights had been completely rebuilt on a more lavish scale than before, and the Community Church had had a new wing constructed, replacing the seriously damaged frame tower.

Norton's plant had been re-roofed. Curtis Apartments and Great Brook Valley Homes had been repaired, but Lincolnwood (slated for demolition before the tornado) had been demolished; a few trailers were still there.

Demolished buildings at the Home Farm had not been replaced. The tornado was followed by a number of investigations and reports by personnel and consultants to the various operating agencies. These post-mortem studies of the effectiveness of counter-measure organizations: Civil Defense, Red Cross, churches, and hospitals in particular – must have resulted in a great many specific alterations and "tightening up" in social organization which are not reflected so much in any physical construction as in disaster plans, inter-agency agreements, and definitions of authority and responsibility, and on these matters I have no report.

Some physical reconstruction may already have come about, however; in this connection I recall noticing that a wider gateway had been made for ambulance traffic at one of the hospitals, and this may have been a response to the experience on the night of June 9, 1953, when many of the hospitals reported traffic jams in ambulance driveways and entrances.

It might be speculated, at this point, that in view of the great concentration on rehabilitation, there may very likely be, after a disaster of this kind, a sort of "undertow" which works against instituting changes rationally recognized as prophylactic against future disasters.

In a general atmosphere of striving with considerable success to reconstruct the past situation, it may be relatively difficult to make innovations. In other words, where rehabilitation is even moderately successful, interest may tend to focus on recreating the old culture rather than on building a less vulnerable culture for the future. And this tendency could, theoretically, actually produce a sort of fixation on the pre-disaster state, which would actively inhibit the institution of changes designed to prevent a similar disaster's happening again. One is reminded here of the building-the-village-again-on-the-side-of-the-volcano reaction. Perhaps persons who have managed to get through the time of maximum stress prefer to repair rather than to prepare (even if repair makes it impossible to prepare).

VIII: SPECIAL TOPICS: The Disaster Syndrome, The Counter-Disaster Syndrome, The Length of the Isolation Period, The Cornucopia Theory

1. The Disaster Syndrome

In my initial field memorandum on the Worcester tornado, and in the later memorandum on the literature on human behavior in extreme situations, I drew attention to what appeared to be a very common behavioral reaction, which had rather definite stages. I called this the "disaster syndrome."

The disaster syndrome is a psychologically determined defensive reaction pattern. During the first stage, the person displaying it appears to the observer to be "dazed," "stunned," "apathetic," "passive," "immobile," or "aimlessly puttering around." This stage presumably varies in duration from person to person, depending on circumstances and individual character, from a few minutes to hours; apparently severely injured people remain "dazed" longer than the uninjured, although this emotionally dazed condition is no doubt often overlaid by wound shock. The second stage is one of extreme suggestibility, altruism, gratitude for help, and anxiousness to perceive that known persons and places have been preserved; personal loss is minimized, concern is for the welfare of family and community. This stage may last for days. In the third stage, there is a mildly euphoric identification with the damaged community, and enthusiastic participation in repair and rehabilitation enterprises; it sometimes appears to observers as if a revival of neighborhood spirit has occurred. In the final stage, the euphoria wears off, and "normally" ambivalent attitudes return, with the expression of criticism and complaints, and awareness of the annoyance of the long-term effects of the disaster. The full course of the syndrome may take several weeks to run.

The frequency of this syndrome, and particularly of its first stage, is difficult to learn. If one used the data presented in Chapter 3, one would conclude that in Worcester, at about impact plus-fifteen minutes (on the average) approximately 33 per cent of uninjured or slightly injured persons were displaying stage one. Probably the incidence of stage one declines in proportion to time elapsed after impact, so that the earliest observers would have seen it more commonly than later observers.

The conditions under which this syndrome occurs seem to be four: an impact which destroys or damages much of the visible cultural environment and kills or injures, or threatens to kill or injure, many people; an impact which is unexpected; an impact which is sudden; and an impact with whose consequences the individual is not trained to cope.

General cultural differences seem not to be a major determinant, since such a "dazed" reaction is reported from widely scattered disaster-struck communities. I would speculate that the more sudden the impact, the more unexpected, the more destructive and the less trained the population is to act in its wake, the more severe and the more widespread will be the "dazed" reaction.

Plate 9: THE DISASTER SYNDROME: SILENCE AND IMMOBILITY

Newspaper photographer's comment was: "She sat. quietly, not speaking."

In this section of the report, I shall try to delineate further some of the characteristics of the syndrome and the conditions under which it occurs, after presenting certain documentary materials relating to it.

A psychiatrist at Worcester State Hospital, and a resident of Curtis Apartments, was in his car during impact. After caring for himself and his family, he went into the Great Brook Valley and other areas to give first aid. His description of the isolation period in one of the worst parts of the total impact area is the best single account, in my files, of the quality of behavior and emotion during the isolation period. I should like to quote from it in extenso:

"... I got into the rubble and by a foundation was an elderly man, there were five people sitting there, an elderly man, say sixty-five, with his wife (I presume) in his arms, her buttocks on the ground and he was holding up the back part of her and bracing her and crying, 'Momma, Momma, Momma' and she was obviously maimed, uh, as I got closer I saw a young man and a young woman, the young woman had a baby, I would say fourteen, sixteen, eighteen-month old baby.

The baby was a little bit bloody and the woman had a... as though she had a huge superficial abrasion on the side of her face. Made her face look, her features were intact but entirely red and raw. I looked at the old woman first and she had a severe compound fracture of the humerus. You could stick your fist into the wound and see the radial artery and what I presume was the radial nerve, something white in there, apparently intact. Although I doubt it, I think she was probably just dazed. She had a head wound, her eyes were closed. She was incoherent, uh, uh, moaning. I, uh... and she kept moving her arm, this was her right arm, and that gave me the creeps, seeing her do that with this horribly broken arm, but she could move the forearm.

I told her husband to lie her down and she was on a hillside, that whole street is. She was head down on the hill, and I, and she was quite obese and she didn't lie well on the ground.

It worried me that she wasn't in a comfortable position but then I thought, well, if she's gonna be in shock, it's better that her head is down anyway and then I thought, well she might start bleeding from her wound but she wasn't. She had some blood smeared on her but she wasn't bleeding; and I *told her husband, and I told her husband, and I must remark that be was one of the few people who seemed to be able to function in a purposeful fashion. Really purposeful.*

I said, 'I'm a doctor, I'll help her, go and get some blankets.' So he did and I looked at the thing and then I looked for something to cover the wound with.

It was full of little bits of glass and it was just a wretched mess, and I couldn't find anything. The curtains were all full of dirt, there weren't big enough pieces of anything to make a tourniquet out of or a bandage; and as I was searching around in a fairly aimless way for something to deal with the wound someway with, he came back with the blankets. So I covered her up, and she was lying on some cardboard or some boards or something, and I thought well it would be best not to turn her over to get blankets underneath her because she's not on the ground and is probably warm enough and she could be made more seriously injured elsewhere.

So, I covered her up. And then I spoke to the young woman, and she said, 'My baby, my baby, is my baby all right?' I looked at the child, and apparently it was. Seemed a little dazed but it didn't have any marks of any wounds on it, just was dirty and had some blood, I think from its mother, on it.

We found an old man who was dazed in the grass in his underwear and stocking feet. I couldn't find a pair of shoes or anything to put on his feet and there were nails and glass and things all around so he and this attendant went off down the hill to get into the street at the foot of the hill to be moved later, at least to get him out of there. I went past his house and there was a woman lying in the only clear space I saw in this dead-looking lawn - all the lawns were killed by this instantly, they looked brown and old. She had a, what I thought was a compound fracture of the femur, the right femur, but I got a progress report on her later and it wasn't fractured, she just had a severe laceration as if a 1 x 6 or a 2 x 4 something like that, had gone right through her leg and passed on. It was a horrible wound, as big as your head, and she was very upset, but did not complain of pain.

Later, I found out she had a fractured pelvis and a fractured, oh, clavicle or ribs or something like that and was bleeding inside, but she didn't complain of pain. But I told her I was a doctor, and I had to get right up close to her and talk right into her ear because she was just moving around aimlessly looking, searching, out some contact more or less.

I took her face in my hands and I said, 'I'm a doctor and I'll try to help you if I can.' And she, the only thing she said was, 'Will it come back? I'm afraid it's going to come back.' And I said, 'No, it's not going to come back.'

I covered up this woman and her husband was standing – let me give you a picture of it a little – of what the situation was and I can place these people. I was rather curious. There was a plot of grass, perhaps fifteen to twenty yards square, and at one side was part of a house, the floor and pieces of partition, off its foundation but sitting there.

There was a woman, who was the daughter of this wounded woman I saw who I'd covered up with blankets (it was the only thing I could think of to do. I was still all alone; the people who had come to help had gone back, I'd sent them away. That thought had never occurred to me before).

This woman, the young woman, was wandering around in the grass not doing anything, and would say something to her mother lying there and then wander away; and I don't remember what she said but it strikes me that it was inconsequential conversation. Her husband was standing on the corner of the floor going through something like uh, some old papers; mind you this was not more than, at the most, the most generous guess was twenty minutes after the thing hit. I don't think it was that long but say twenty minutes. He was going through some papers (my fantasy is that he was humming while going through these papers; I don't think he actually was, but he was very much unconcerned) and then it started to rain, and God damn it, all these people get wet, what could be more miserable? So I said to him, 'I don't have enough covering over your wife, she's badly hurt, get me some blankets, you must know where there's some in the house.' Well, there was none in the house, it was all outside and he tried to lift a partition off the bed and get some and he was very reckless about it because the thing was too damn heavy – nobody could lift it off; but he kept trying in a very desultory fashion.

Closet wall was off and I just reached in and I said, 'I'm going to take some of these coats and cover her up,' and he said, 'Oh, yeah, you go ahead and do that;' and then he stopped me and said, 'Don't take my good coat.' And I had the feeling that he was like a crazy person, not in good contact or whatever words would describe it, childlike; and I said, 'All right, I won't,' and I took what coats I needed anyway and covered her up.

Then George, the attendant, came back and we went up, I told him probably the best thing to do was get the walking wounded people down as fast as he could, just get them out of there in a group; and he proceeded to do that and I presume he did for quite a while, because there were a lot of people obviously superficially hurt but bad enough to need medical attention, and he was taking them down because they were just either sitting or walking around or doing nothing.

There was a young man there I would judge twenty-four, and he looked very business-like and sort of preoccupied and looked at me like, 'What the hell are you doing here? I'm taking care of things;' not exactly hostile, but self-sufficient. And I said, 'I'm a doctor. Is anyone hurt here?' And he said, 'Well, my mother's under the house.' That house was off the foundation by ten yards and I looked under the house and this woman, it looked like the whole house was sitting right on her chest. There was some blood on her mouth and I thought perhaps she was bleeding from her lungs. She was taking very short breaths and had very little breathing space.

He had gotten an automobile jack and put it under the house and jacked it up as hard as he could to keep the house from going any further down, but he couldn't lift the house. Then I found George and we tried to, we thought perhaps it might teeter, and we tried to lift up and it wouldn't, and we tried to pry and it wouldn't pry up with a long lever; and so then I decided perhaps we should look around for some more automobile jacks and see if several couldn't lift up the house enough that we could get her out. I didn't want to crawl under there.

Uh, then I heard sirens and I felt relieved because they were obviously coming to our area and we would have some help. So, George was back and we decided to get some more automobile jacks and had a hard time finding them because the cars were, they were sprung, you know. There were lots of them lying all over, but we couldn't get 'em out. And in one of those cars was a fellow who behaved very curiously.

The car was on its side, and he was sitting in the car on the back seat with his foot out the back door, and the weight of the car was closing the door, of course, and his foot was outside. His friend was standing there and called me over when he heard I was a doctor and wanted to know what could be done. I got in the car. I had them hold the car because it was teetering and I got in and looked, and as far as I could tell the leg wasn't broken and he was, the guy was complaining of no pain. It's curious though that he had a gun beside him. I didn't like this but George did, and George had the fantasy that he was going to shoot himself; uh, I just say this because perhaps George's observation was more accurate than mine, but as far as – I don't really remember the fellow except that he seemed to be very cool, that is, in his talking. He was obviously tense, and quite pale but he was very cool. He said, 'Don't tip over the damn car. It doesn't hurt. It throbs a little bit but let's just leave it there. I don't know what the hell will happen if you tip over the car.' So, I got out a jack and gave it to somebody, I forget who it was, and said to go back and put it under that house; and then I crawled underneath the car, by the trunk, and could see that the foot, and again I didn't think it, there was no bleeding and I didn't think it was broken, and looked at the car and decided we could tip it over if we tipped it over easy. So several of us got together and tipped it over a couple of feet, and using a pry, himself, a piece of stick, pried open the door enough to get his foot out. It didn't seem to be broken. But he was very relieved and then became very – uncool. The climax was over and he was very, well, 'I don't want to do another damn thing, I'm all tired out, I'm drained of energy,' and things like that he said, and he sat down on a rock.

On the lawn was a boy about twelve years old and his father was sitting by him. His head and upper chest was swathed in bath towels and he had a cut, very deep, extending from the vertex going down behind the ear and into the neck down into the chest. It was about a foot deep and laid open a *flap almost to the right orbit, and he was twisting and jerking and I thought perhaps he might have a cord injury or that he was dying. I lifted up the towels and looked and felt that I could do no better, wrapped him a little tighter, a little pressure, a lot of it was necessary; and then I told his father, I looked up on the hill where there didn't seem too damaged houses and I said, 'Your boy is terribly hurt, but not dead. Go up to the top, there where those people aren't hurt.' And he left, and I got the kid's brother, who was sixteen, I would say, to sit by him and watch him if he started to bleed or anything like that. And then I went around the corner and here was this kid's father, talking with somebody in a friendly, neighborly sort of way, reminiscing or something; and I said to him, 'Can you go?' And he looked at me almost breaking down and said, 'No, I just can't leave, I just can't bear it.' So I said, 'You go back with your son and I'll go, as soon as possible I'll see that. somebody comes for him.'*

As I would encounter people who were not hurt I would say, 'Help me move some of these wounded people out,' and they couldn't. I would talk with them and explain what I had in mind, and they seemed to react and respond in a normal way, and then just stand there or wander away, and nod their heads and not do anything. I wouldn't say that they were dazed, but they were not functioning; and almost everyone I met was like that. And as I look back, my own behavior shows this, although I wasn't aware of it at the time.

I was beginning to get tired. I don't know how long all this took, it seemed like about a half hour, but it must have been longer than that. Must have been an hour or more. I went back through the rooms, and helped carry some people down. I asked a man if anybody was hurt in his house, and he pointed to his dead wife on the ground and said, 'She's hurt,' and I said, 'No, I mean somebody I can help,' and he shrugged and said, 'no', he guessed not; and then he went back into the rooms of his house to pick over some things; he seemed to be unmoved by the fact that his wife was dead.

Many people, when told I was a doctor, became very dependent on me, uh, 'What shall we do, where shall we go,' but then couldn't do it; and other people didn't seem to be able to take in the fact that I might be able to help them. Wouldn't answer me.

I would say, 'Is anyone hurt here?' And they wouldn't answer..."

The impression to be gained from the psychiatrist's account agrees with the estimate, made earlier on the basis of interviews with victims themselves, that a large number of the persons in the impact area during the isolation period (i.e., up to about half an hour after impact) were incapable of more than minimal (and often inadequate) care for themselves and their families. Many of these uninjured apparently were unable or unwilling even to walk out of the area.

Far from panic, there was a dissociated state variously alluded to as "daze," "shock," "apathy," in which the normal cognitive structure was severely limited and effect was limited except for occasional emergences of irrational dependency, hostility, or "hysteria." One has the feeling that the dazed, helpless, floundering state represents a partial regression to a level of behavioral organization in which the individual is limited in his ability to perceive the actual physical environment and cannot act coherently with regard to the objects and dangers around him. We shall discuss later possible determinants of this regression.

Plate 10: THE DISASTER SYNDROME: THE "DAZED" REACTION

The photographer's comment: *"Emergency treatment being given one of the tornado victims in wreckage of her own house near Great Brook Valley, where she was found sitting bewildered in the debris."*

The first rescuers from the community aid area found the survivors in a random movement state, similar to that described above by the psychiatrist. In the words of a fireman in the Home Farm and Great Brook Valley area:

A. ...Then we saw the people that were walking around there in a daze. They didn't know what hit them.

Q. You mean, just walking?

A. Just walking around as if the world had ended, you know. Some of them, you could see that look in their eyes, didn't know what hit them, see...

After recognition that help had arrived, however, the dazed and apathetic stage disappeared, and survivors participated more extensively in rescue, first aid, and evacuation activities. Another fireman described both the "random movement" and the probably following "helpful" stage as his fire engine moved through the Burncoat Street area on a circuitous route toward a fire on Francis Street:

...There were quite a few people around, but they weren't in any particular spot; few here, a few over there. And everyone had, although not exactly a stunned attitude, a shocked attitude. Some, enough to realize that there was somebody in that house and needed help to get out. Others just seemed to, well, watch you go by; probably completely shocked for the time being...

...when the officer told us that we were to lay our lines down to the pumper on lower Francis Street, I don't know where all the help came from...

...that thousand feet of hose went down that street just about as fast as you could let it run; and there a couple that hollered, "Come on, give us a hand with the line." And just as fast as one man could pull it out of the wagon, it was goin' away from us. It was amazing. In fact, I've never seen a hose line move quite as fast as that one did. The people around there were very cooperative in that case, very cooperative. And some of them, I imagine, from that house where we were trying to get the little girl out, to help us. But there was more on the line than there was over at the house, so they came from other places too...

The pastor at St. Michael's on the Heights Episcopalian church, was away from the impact area at impact, and was able to return by 6:30, during the height of the rescue period. By this time, it would seem, a common modality of behavior was mutual reassurance. Finding his family unhurt (the house was in the fringe impact area), he donned his cassock and made the rounds of his parishioners' homes asking, "Are you all right?"

"People would cry when I asked." "I feel they knew I represented God as well as my own personal concern for their safety." "People wanted more than anything else to know people cared about them." "If they knew me, they would give me a hug. They wanted to be reminded that they were all right." "They wanted comfort rather than physical help."

Another minister who was also making the rounds of his parish about the same time remarked, "Almost everyone wanted to go beyond a handshake, wanted an embrace; wanted to lean on you. Some want to kiss and be kissed in reassurance." Both ministers referred to the cooperativeness, politeness, and willingness to help which the people showed to each other in the impact area during the rescue period.

Among those being evacuated to the hospitals, a similar type of complaisant, helpful, and grateful attitude was observed. A series of interviews by Miss Rayner with medical personnel, who were acting independently at the hospitals during the disaster, convey a remarkable uniform picture:

A neuro-psychiatrist who saw victims at Memorial Hospital from 5:15 PM to 2:00 AM:

A. ...I am not sure whether this was apathy or not; it seemed more as though these people felt themselves a very small part of some one great thing. It was not a real apathy. People were extremely patient - not demanding, and seemed not to be focused on themselves. The only outcries of pain were patients with fractures who had to be transferred. They did respond to pain, a few to the needle, but not as many as I had expected. I was impressed most by the quiet. Responses were less than expected. There was real concern for family and house. I saw one man quietly crying to himself – he had a fractured leg – when I asked why he was crying, he told me that his little girl had been killed. A twelve-year old kid asked to have others cared for first...

Q. Had they an awareness of the environment?
A. Yes, but the severity of the experience has numbed them down. I have my own theory; it's related to Selye's work - trauma like nothing they had ever experienced before.

Q. Did you overhear any of their conversations?
A. Victims by and large were not talking. The medical personnel did all the talking. My usual technique of kidding and trying to cheer people didn't work. These people did not respond. Children were helping (kept them occupied) and very grave –behaving like "miniature adults," no fooling among themselves...

Mrs. (**N**ame **W**ithheld): Much more confusion – more feelings expressed. The expression was not adequate here in this situation. Most were oriented, no anger expressed or that type of emotion, characterized by "why did it happen to me." There were no complaints.

A. They seemed at once lost in a vast human field. They were not like a regular accident victim. A day later they were in a dazed state. There were no complaints – no kidding – no horse-play by youngsters. It was such a part of such a big thing, there was a transient loss of individuality and identification with something else.

Q. Did this appear as withdrawal?

A. It was withdrawal from the self to something bigger than themselves. A vast investment of interest in others. No one came to office without referring to tornado.

Mrs. (**NW**): There were no evidences of selfishness at all. There seemed a high threshold for pain. Separation seemed to be the most traumatic to them.

A. A few were confused – dazed and "blank." Mothers were going through wards at City Hospital looking for children, and acting "like animals."

Mrs. (**NW**): They bad a one track mind. The nearest to panic occurred when others began looking for family. There was an urgency and desperation.

The Superintendent of St. Vincent's Hospital:

Q. How did the patients look?
A. Stunned, but very cooperative and willing to do whatever doctors told them.

A Nursing Adviser and Coordinator:

A. People so shocked they didn't know what had happened. They were so stunned, their response to local anesthesia – was even not where it needed to be used. When you talked to victims they didn't seem to respond.

...Wednesday at 9:30 A. M. the Red Cross took over first aid. Hospitals were still taking patients. People didn't realize that they were injured until several hours after the tornado.

In an Interview at Memorial Hospital:

Dr. C: There were an extreme number of patients in shock (wound shock).

Q. Did you notice an element of emotional shock?

Mrs. K: ...The relations were in a state of apathy; it made them easy to handle. One grandmother had lost two children; she showed little response.

Dr. C: One man simply stated that his wife was dead; there was no emotional expression.

Plate 11: THE DISASTER SYNDROME: BODY CONTACT

The Superintendent, Worcester City Hospital:

They were dazed individuals, many unconscious, many cut, 40 per cent fractures of long bones, head and chest injuries. Most were so Dazed they felt no pain. Parents were disturbed over children.

The Director, Hahnemann Hospital:

Q. How did the patients seem?

A. No hysteria – they were stunned – just came in and were willing to do as asked.

Q. How did they look?

A. Dirty! Just covered with mud. Gravel was ground into their scalps and skin.

Q. How did they respond when spoken to?

A. Oddly enough, most were well oriented. It was not until you listened to their stories that they reacted. They seemed to feel lucky that they were alive. Great appreciation of what has been done for them. We got them to homes and friends. All knew of someone who would take them. Only a few had to have temporary shelter away from friends and family.

The Superintendent of Worcester State Hospital:

A. I went directly to the housing project (Curtis Apartments) in an ambulance from Worcester State Hospital. Then went to Memorial Hospital and worked. In actually only about a half hour later, Dr. Hicox was there. My feeling regarding the thing – my interpretation of those actually in the area, especially those brought to the hospital was quite surprising. People actually in the area were calmest of all.

Q. What kind of calm?

A. Most of a pathetic, lethargic, somewhat stuporous type. (He then told an anecdote about a woman patient) It was as though she were almost in shock, yet still standing. Unemotional, complete flat effect, nothing seemed to mean much to her. One woman was standing with a piece of cloth in her hand in a ruined house. I asked her if she were hurt – said no – she refused to leave home. I found that she was hurt. There were no complaints. Some were completely unresponsive. At the hospital no one was hysterical, complaining or crying. They were quiet and stayed put. The people working, nurses, Red Cross, attendants, volunteers, nurses-aides, those groups, were excited – rushed about here and there, causing commotion and interference, and emotionally were all upset. Some of the nurse's aides even would start out for something and never get back. There were a lot of patients but no complaints. One woman claimed she was unhurt, that she had no pain; said she had a muscle strain. A doctor examined her and she had six ribs fractured. Had to demand to examine them; most denied pain. One man had both hands fractured – this patient disclaimed injury – said he was waiting for his wife. The doctor set the fractures and the patient rolled over and went to sleep. He didn't care. Those in it, it was such a shock, that they were insensitive to sensation or demands. They were admitted, but most didn't know quite how they had gotten there.

Q. Were there any really disoriented?

A. Some were partially so. One woman had three kids with her, aged eight, nine, ten. She knew and asked me to take care of the children. She didn't know she had been hurt, or how she had gotten to the hospital. She said she was going back to the house to get her husband and a shoe for one child. She didn't realize that one child was missing. When I mentioned the missing child, she wanted then to go to find the child. I had volunteers take her and the children to the Auditorium. She wasn't upset; she didn't cry; she was not hysterical, and stuck by the kids. I was astonished at the general public – wonderful, but helped in a confused way. Wanted to help but didn't know how. There was not enough supervision. No one would stick with you, doctors and nurses were the only ones who would keep their heads. The volunteers interfered with the functioning of the hospital.

Q. Was there apathy all about?

A. If these people were not shocked they'd be very difficult to handle; because they were shocked you could handle them easily. It probably helped them through the immediate post-impact period. I was most impressed by their lack of pain. No patients (mentally ill) were received at State Hospital as a result of the tornado. Some of the patients at the hospital lost family (two schizophrenic patients). They had no reaction to news. You could do anything to the patients (victims), but could get no response. Those who came into the areas (sightseers) and those volunteers in hospitals were shocked by what they saw, but couldn't seem to do anything.

A Civil Defense secretary:

Q. How did people at the Municipal Auditorium aid station look?

A. Right up tight! The elderly were very calm, seemed almost to enjoy it; the kids played as usual. The men, well one man seemed to blame himself, "shouldn't have lived there." No emotion, sort of "deadened." One girl (from a well-to-do family) just went to pieces. She didn't seem to care; then when we found her, she cried.

The medical report by Bakst et al, 1953, included the following comment:

"…During the first night of the disaster emotional reactions requiring special attention were rare in the hospital; most observers were impressed with the composure and docility of the casualties. Indeed, some patients were in a state of apathy so extreme that serious injuries did not seem to be of concern to them. Even the injured children were quiet, obedient, and anxious to be helpful…"

These observations on victims in the impact area and in the hospitals during the rescue period show a partial continuation of the dazed state, with relatively flat affect, but an improved capacity for action if organized community aid personnel (physicians, nurses, firemen, police, et al.) were present to give directions. In the presence of such personnel, with equipment, giving instructions, the victims were markedly obedient. They also indicated intense gratitude and pleasure at receiving any help or verbal expressions of concern or interest.

Injured victims seem usually not to have gone into wound shock until after evacuation; but the incidence of wound shock after evacuation was apparently high. The interplay of physical and emotional trauma in producing the "disaster syndrome" should be investigated.

The next stage, the early rehabilitation period, saw a further development of the syndrome. In this stage, a cardinal feature was altruism. The dazed component and the extre.me passivity had passed. In the words of the Bakst report (p. 28). quoting Bishop Wright:

"…The general impression of observers is that the injured and uninjured alike were more concerned for others than for themselves and were so awed by the enormity of the disaster that 'each person became a saint for about ten days'…"

One Protestant pastor whom I interviewed also testified to the prevalence of the altruistic component among the residents of his part of the impact area up to June 19. This pastor noted that among his parishioners, in the Burncoat area, there was a real "revival" of community and church spirit; there was much freer personal contact; manners were more "natural;" everyone pitched in to help clean up the church property; old feuds were forgotten, and persons with mental disease symptomatology or histories of hospitalization showed a notable remission of symptoms or at least no relapse. Another pastor commented on the continuous desire victims felt for people to come to them, their anxiety and fear at having to go to apply (e.g., to the Red Cross). Perhaps the altruistic component has a demanding and dependent side as well as a generous one.

Consolidating the data given above, one can describe the overt behavior of the disaster syndrome as displaying three stages, corresponding roughly to the isolation, rescue, and early rehabilitation periods:

1. Isolation period: many survivors in the impact area, injured and uninjured, are "dazed," "apathetic" "stunned;" awareness of the extent and intensity of destruction, to person, family and community, is inadequate; efforts at first aid. rescue, and evacuation are perfunctory, frequently inadequate, and often entirely absent; many people simply stare, wander aimlessly, "putter about;" expressions of strong emotion (grief, fear, pain, anger, etc.) are missing or sporadic and inappropriate.

2. Rescue period: with the arrival of organized protective personnel, with equipment (firemen, police, Civil Defense, Red Cross, etc.). survivors remaining in the impact area are able to perform routine tasks in rescue, fire-fighting, etc., operations, under orders; survivors evacuated to hospitals, aid stations, and other mass care centers are also able to follow orders and are extremely docile and obedient, but they tend to remain "dazed" longer, or, if injured, to go into severe wound shock; both groups are better oriented than during the isolation period and can usually tell an interviewer what happened to them; both groups are extremely grateful for care and concern; both groups are also self-sacrificing and willing to let others be cared for first.

3. Rehabilitation period (first ten days at least): both injured and uninjured, in hospitals and in the impact area (but perhaps not to the same degree if they are no longer receiving evidences of mass care, as in cases where evacuees have gone to live with relatives) show a mild euphoria, marked by intense altruism and willingness to work for the community welfare, readiness to give up old grudges, to ignore barriers of social distance, and to merge the self in a kind of neighborhood revival spirit, to participate in a strong "we-feeling" shared by impact-area survivors but not by others; at the same time, in spite of professions of thankfulness and gratitude, there is also a tendency, inconsistent with the foregoing, to complain of the coldness and efficiency of mass care organizations, to become very sensitive over applying for aid in any form, and to be willing to accept assistance only if it is brought to them.

The three stages of the syndrome might be labelled, for purposes of reference, the random movement stage, the suggestible stage, and the euphoric stage. (In the earlier description, a fourth stage – an ambivalent stage – was differentiated from stage three.)

This syndrome describes the behavioral modality of an unknown proportion of the survivors in the impact area. Some persons – notably those with pre-defined roles to play in a disaster – do not succumb to the syndrome, although they probably have tendencies toward it; some few persons seem to escape it by immediate and direct "hysterical" outbursts; some, of course, are so severely injured that physiological trauma governs behavior. (The incidence and nature of the "hysteria" occasionally but vaguely described is another point worthy of investigation.) I would hypothesize that this syndrome will invariably occur following a disaster characterized by a sudden impact involving physical destruction and injury affecting a large part of the survivors' visible community environment, and that cultural differences will not affect it appreciably[35]

The determinants of the syndrome are, in my opinion, not primarily physical injuries, physical shock, physical upset by buffeting and being "pushed around by a lot of air;" they are psychological. This is evidenced by the fact that the disaster victim can display this syndrome whether or not he has been injured; it is also

---

[35] Descriptions of "typical" behavior at Hiroshima and Nagasaki and in severely bombed German cities, after earthquakes, and after sudden floods, while not organized as here, seem to be referring to the essential elements of the disaster syndrome.

evidenced by the fact that injured disaster victims display the syndrome more frequently than do accident cases (the latter being said usually to be noisier, more demanding, more sensitive to pain, more susceptible to the physician's bedside manner, etc.).

Plate 12: THE DISASTER SYNDROME: PASSIVITY
A victim from Great Brook Valley is being registered for transportation.

The precipitating factor in the disaster syndrome seems to be a perception: the perception that not only the person himself, his relatives, and his immediate property (house, car, clothing, etc.) have been threatened or injured, but that practically the entire visible community is in ruins. The sight of a ruined community, with houses, churches, trees, stores, and everything wrecked, is apparently often consciously or unconsciously interpreted as a destruction of the whole world. Many persons indeed, actually, were conscious of, and reported, this perception in interviews, remarking that the thought had crossed their minds that "this was the end of the world," "an atom bomb had dropped," "the universe had been destroyed," "the whole city of Worcester may have been destroyed," etc. The objects with which he has identification, and to which his behavior is normally tuned, have been removed. He has been suddenly shorn of much of the support and assistance of a culture and a society upon which he depends and from which he draws sustenance; he has been deprived of the instrumentalities by which he has manipulated his environment; he has been, in effect, castrated, rendered impotent, separated from all sources of support, and left naked and alone, without a sense of his own identity, in a "terrifying wilderness of ruins."

The response to the assault of this realization is withdrawal from perceptual contact with this grim reality and regression to an almost infantile level of adaptive behavior characterized by random movement, relative incapacity to evaluate danger or to institute protective action, inability to concentrate attention, to remember, or to follow instructions. Such individuals appear to be "dazed," "shocked," "stunned," "apathetic." Actually they are far from being indifferent; it is the intensity of the previously-felt anxiety which has prompted this blocking of perception and this regression.

The remainder of the syndrome represents the gradual restitution of the pre-impact behavioral organization. The individual first seems to emerge from the cocoon of apathy.at the appearance of rescue personnel from the outside; he becomes very dependent on these people, identifies with them, is extremely suggestible and obedient, idealizes the care they give, rejoices in embracing and touching them. <u>This is the suggestible stage.</u> Following this comes a stage in which there is a joyful resurgence of identification with the community, a reacceptance of the adult role, although the reassurance of unsolicited aid and assurance is constantly demanded. Finally, there is the development of complaints and blame. This is an ambivalent stage. Eventually, no doubt, most individuals return approximately to their pre-disaster state, after perhaps two weeks or a month. The various phases of the syndrome, after the initial plummeting regression, appear like a telescoped passage through the familiar stages of behavioral maturation.[xxvi]

The most transparent case history of the first two stages of the disaster syndrome is provided by an interview obtained with a middle-aged small-businessman who weathered the impact in his truck. After impact, he crawled out of the truck and got back into his place of business; it was a shambles, but the phone was still intact.

A. ...I came back in here and I called up a friend of mine up at the Brookside Home and I said, "My old truck blew down - can you give me a lift?" And he said, "I'm eating my supper and I'll be down." So, I thought I was, that this was just something that happened to me, that was, you know, the old shanties here could blow down any time, you might feel. I called up this friend of mine, he says, "Everything gone?" And I say, "The whole works blew out." He said, "I'll be down after supper." Well, you know, my damage wasn't as teetotal as what happened to a lot of people so I don't bother around here (laugh) or anything. I couldn't quite convince him that something had happened. He didn't get here until well, past eleven.

Q. Tell me, when was the first time you realized that something was really going on that was really wrong?
A. I think when I was on my hands and knees.

Q. Didn't you realize when the desk went out in front of you?
A. No, there was so much...oh, suppose that you were sitting here... and a tidal wave flowed over you, you were just struck. I think - that you struggle before you think, you know, you don't plan anything at all.

Q. So, it was about when you got on your hands and knees that you really felt something was...?
A. Something was bad wrong here – but I'd never had any, well, I'd read in the papers and so forth, you know about tornadoes, and about terrible disasters high winds had caused everywhere else; and I was right here, uh; I lived across the street to here during the hurricane and the hurricane wasn't as terrible as this. And I was only on the fringe; it's a lot worse, two minutes' walk from here. Remember I had my head down and my nose down; I was taking care of my eggs, and my chicks, and making for the truck – I felt if I could get

in the truck... I'm not sure whether I can remember it – if part of the roof had gone by then or not. The noise was, terrific. But on a tar paper roof here in a hail storm like that the noise is terrific too – what you would expect, a good steady roar there for a couple of minutes before I got out of here. So, although a whole lot of people can give you a detailed description of how the storm looked as it approached, and all that – I think if they were busy – and of course after the storm got here they would be busy – I can't see how people could see it at all. I've been over in the Adirondacks, washing dishes when I went to school, wash dishes up in summer resorts, summers; and they used to have terrible hail storms over there, cloud bursts and thunder storms, worse than I've seen around here. And I do think that perhaps this, as far as I knew it was just a bad storm.

Q. Then when you got in the cab of the car, were you looking up and seeing what was going on? Or what was going through your mind then?
A. No, I looked at uh, 'course through the windshield I could see that other building there, and there was a lot of stuff in the air; but the air was thick with stuff. Your vision wasn't far, or anything like that; the air was actually thick with debris, shingles, dust, and dirt. Kind of a red powder. Gray or reddish powder. That's as much as I remember.

Q. And you say when you got on your hands and knees, you felt that this was something happening just to you? What did you think it was?
A. Well, I didn't uh, it was just a bad time (laugh) I was getting out of it if I could.

Q. Did you have any idea what it might be due to, I mean, did you have some thoughts about it?
A. No, I was a lot scarier afterward when I walked up the street and saw, you know, uh carrying bodies out and so forth. Until that time, I didn't realize the seriousness of the storm except that it had blown hard and like I said we are educated here to how much damage a tornado can do. So, if the wind blows hard here, it blows a gale, we have a thunderstorm; that's all I thought we'd have here.

Q. You say you didn't feel scared really until...
A. I think, you know, after the truck came back up, and still kept swinging and heaving in the wind, and trembling – I kept, I'd look up through the windshield and think I shouldn't have got this truck (laughing) I should have kept the old Dodge I had. The windshield was too wide on there (laughing) and it looked terribly big to me with the stuff that was in the air – shingles, roofs and stuff like that swingin' by there, that is, you could see something come by you, you couldn't actually identify it.

Q. Did you notice any feelings inside yourself at any time during this?
A. No, (loudly and slowly) I don't think that I noticed any feelings, although people came in afterwards said they thought (chuckling) I must have been stunned a little bit.

Q. Why did they say that?
A. Oh, they thought I should have got going a little quicker, gettin' a hold of the insurance man and stuff like that. 'Course I thought at the time I was doing my utmost and no one could do any better. (nearly inaudible).

Q. Looking back on it now what do you think about it?
A. I think I might have been sort of... riled (low voice) because I, uh, after the...died down, was gone, my phone was working for ten minutes or so, I called up this fellow and asked him to come down and give me a lift and straighten this place out down here, not realizing the extent or the seriousness of the storm. I thought this was just something that happened to me. I called up my wife and she said, "Well, there's a couple of trees gone in the back yard, but that's all." So I said, "Good for you, sit tight, where's Ann?" That's my daughter and she's working over here at Holmes, that's the greenhouse. They were hit bad over there. So, I left with the truck and I went up to as close as I could get to Holmes, and I couldn't get up there. There were trees lying down across the road, so I came back and went down to the drug store and tried to call Holmes but the wires were down. I came back here and left the truck and walked over to Holmes. Some way or another I missed my daughter, she was down here, looking for me. But the people in

the house there told her I was all right, that's as much as I know about the storm.

Q. You say maybe you were a little slowed down, I wonder if something like this happened again, heaven forbid that happening, but do you think you could be more efficient in the next one or do you think you'd do about the same?

A. Yeah, I think that in another one perhaps I could think, you know, I wasted a lot of time here. I couldn't do anything myself, here. I could have gone out and given some other people a lift, if I had, uh, if another wind blew like that I would know that someone possibly could be hurt or dying on the next street, and I could go help them. But, uh, and I think I could have even with the truck and myself.

Q. But thinking back on what you actually did how do you feel about it?

A. I think I looked after myself. And, of course, the first half hour or so I thought it was something that only happened to me. Finally, after an hour or so, my father lives up on Garfield Street; I walked up to see if he had been home, and if he was, how he was. But he wasn't home, and I made certain of that, and that he was safe. But I think in another storm I would know a little wind can do a whale of a lot of damage; and I'd get right out there.

Q. But you think that not knowing any more than you did, and the situation being what it is, how do you feel about the way you carried things off?

A. Oh, I think I did all right but maybe I was a little dopey; maybe I still could have got over there. I didn't know for a couple of days afterward that these houses up there on the hill were tumbled down and moved off their foundations. It didn't dawn on me at all. Of course, at the same time I live in this section and we had no telephone, most of the time no newspaper, and no electric lights and., of course, no radio and no communication at all with the outside excepting with people that you talked with. I think that facilities should be set up in a mobile form that could be moved in, and take over telephone communication on each street, or something like that. Because the people in a devastated area can't call out. People outside that are looking for them or might help them can't call in, and do not know the extent of damage or anything. I think actually that partially was why I didn't get out and do my best to help out. Because I was perfectly able to do it, perfectly able.

Plate 13: THE DISASTER SYNDROME: THE "STARING" REACTION

The scene is at Burncoat and Fales Street. The photographer's comment:
"The debris was so mixed that it was difficult to immediately determine how many homes were razed."

Q. Did nobody, nobody came to help you then until 11:00 o'clock that night?
A. That's right.

Q. Nobody showed up around here, rescue workers?
A. Of course, first I went over to Frank Adams over here to tell him my troubles and I see his roof is falling off, and so forth, an' they're sweeping the street; so I said well, I said I'd better go back and sweep my place. So I got back and gee, you know I worked my trade in that building over there, I worked in that building twenty-one years; and when I got back here it caught fire and was burning. So I walked back down there but there was nothing I could do. The fire apparatus had gotten there, there was nothing to do but sit here and watch that old shebang burns down. Well, it burned, you know, it gutted it pretty well. This is the building that…

Q. The building that was burning was not your building?
A. No, it wasn't my building; but figured it was home, once upon a time, practically.

Q. You say there was something here that blew over?
A. The building that blew up that the car was parked against, which is that one. I didn't even see it go but I went that way. It was right here.

Q. Oh I see. The foundation that's left there was its foundation?
A. That's all, there's nothing but the foundation there now. Took most of Murphy's heart, I guess.

Q. What?
A. Most of Murphy's heart. There was the foundation – $9,000.00 worth of stock – which, of course', is nothing if you didn't have your house blow down and lose some of your family.

Q. But it's' still pretty hard.
A. Yes, this was hard-earned, you know? (earnestly) The wood-working industry never was highly paid. I served a long apprenticeship – and during the time that I worked at it, although I liked the trade, and the work, and the people I worked with, I never was able to accumulate two cents during that time. Later when old man Young died, the place was sold out, liquidated, the estate was: I went to work at Norton Company and, of course, as long as you're willing and able at the Norton Company you get a good break. So I was able to lay by enough to start myself, very humbly, in business in the old trade. So hated to see it go. But, like I say, I am still thinking of myself like when I was (chuckle) crawling out there…

Q. …at any time during this thing did you notice any physical symptoms?
A. I don't think so.

Q. Did you ever notice your heart beating fast, your mouth getting dry?
A. No, not a bit.

Q. Haven't had any diarrhea or constipation?
A. Not a bit.

Q. No pains in your stomach?
A. Excepting that I can't sleep.

Q. You can't?
A. No.

Q. What about that?
A. I don't know, of course, it might be these kids (laugh) we have home the wife is taking care of but I had the…the best night's sleep I had was four hours.

Q. Do you have nightmares?
A. No. I think I dreamt something last night but uh…

Q. What do you dream about?

A. (Pause) Dreams are kind of vague. (Vigorously) you know one time when I was a kid, I went to kindergarten over here at Carter and Fairhaven Road. And during recess I climbed the apple tree and slipped down and got my foot caught in the crotch. And you know I can remember, well, the kindergarten teacher, Miss Bancroft came out and unlaced my shoe and yanked my foot out, after I didn't answer the bell. I suppose at the time I figured that was about as bad a thing as had happened to me. You know, last night I dreamt something about Miss Bancroft bringing me back in school. Something odd like that. (Pause). That's about the…That's about the worst thing that had ever happened to you up until then. I think I was terror stricken, you know, when my foot was caught in the crotch of that tree and I couldn't move. And, I don't know, I suppose that when she unlaced that shoe and yanked my foot out of there, and then got my shoe back out of the crotch of that tree, that was security and safety. But there was something in what I dreamt last night about Miss Bancroft.

Q. No more than four hours sleep since this thing happened. How do you feel now?
A. Not too bad. I'm fairly efficient today. I've made out forty-eight invoices, and, of course, I've got to send, uh, I'm planning out some kind of a note that I can send to everybody that, all that buy my stuff here; that if they know that they owe me, would they please let me know how much, I have no record at all. I've got $5,700 bucks on the books and I don't know for certain where it is.

Q. (Whistle) You really have to trust people.
A. I've never worked in anything but dungarees all my life. (laugh). Still it isn't as bad as what I've seen happen around here.

Q. How's your appetite been?
A. Well, I didn't eat much supper or any dinner last night, and I had a cup of coffee and a bun for breakfast yesterday. Uh, today I didn't eat any breakfast but I ate a good dinner down at Stewart's diner.

Q. Is today the first time you've eaten well since…
A. Yeah. Yesterday I bought a steak out but I couldn't eat it. It was the finest looking steak (laugh) I've seen in a long time. And I don't know why; because I can eat one of those good big steaks on hand down at Durgan and Parks, down there.

Q. (Laugh) If you can eat one of those you're a good steak eater.
A. That's right and. it was a crime to leave that steak.

Q. Right after the thing hit you weren't hungry then - you didn't eat that right?
A. No, I didn't that night. No. (Pause ten seconds) I didn't, I went home that night and came back here, then I didn't go home for a couple of days. When I went home that night, 'course the electric stove was off and the refrigerator and all. My wife said, "Drink the milk up, any way, if you can't eat, if you didn't feel like eating." I said I didn't feel like eating. So, I had two glasses of chocolate milk and came back here.

Q. You stayed here, you didn't go back home?
A. No, I didn't go back home for a couple of days.

Q. You just stayed right here and worked?
A. Yeah. The rubble and debris were tremendous. So I took twenty-three truckloads of stuff out of here before we could even see where my material was, where the machinery was and anything else.

Q. Did you notice anything else, or have you noticed anything else?
A. Only that, you know, I'm kind of a happy cuss most of the time and l do have… well I haven't today but ordinarily – you know I've had a couple of hours a day that I feel the worst for myself. I could just sit back and sympathize with myself, and what had happened to me. I've had bad things happen to me, maybe not as bad as this, but nothing ever got me down. Because I've been broke before (unsteady laugh) but it shouldn't bother me too much (nearly inaudible) there's other things to think of.

Q. You just felt sort of like crying?

A. Yeah.

Q. Have you ever been able to let go and cry about this?
A. No.

Q. You don't do that sort of thing?
A. No. Come pretty near it though. (briskly) Fellow came in here that I had worked with years ago and he works at Martins now and I heard that his house was gone. He lives over on Randall Street. They bought the house years ago, oh, during the depression; borrowed everything they put into it at the time; but since then, of course, he's come up in the world. But they're tremendous working people, good people with a tremendous capacity for work and know how to work and how to do things and spent all their money. You couldn't duplicate the house over there for $30,000. And it's a total ruin; and they may get $6,000 or $7,000 out of their wind damage. Well, that guy came in here and l could feel for him. That's as close as I came to bawlin' during the time. But he's got a lot of guts, and what I heard last was he's going to put a trailer up on the next lot and after the house is demolished he'll rebuild. That really takes somethin' because when he does rebuild, now, there was a house that was free and clear: he's going to assume a tremendous mortgage, and he isn't a young man any more. It's all right for someone like, if you figure you have twenty-five or thirty years of good productive years ahead of you, you can get out of anything, anything under the sun. But if someone gets along to be about fifty-five or fifty-eight, something like that, it's a different story. In their lifetime they can't, I can't see how they can put themselves back where they were.

Q. How old are you now?
A. I'm forty-three

Q. You've got a little while yet.
A. I've got a little while but I have a lot of heavy labored years in back of me. Really heavy years in back of me. I haven't got the pep I used to have, the get up and go, and I know I haven't. If I need something done that's got to have get up and go to it, I hire it done. And… I know I'm not as good a man as I used to be.

Q. Tell me, have you noticed any tremulousness, any shakiness?
A. Yes, I have been shaky, yea.

Q. When did you first begin to notice that you were shaky?
A. Oh, a couple days ago. I had a sliver in my hand and I went to take it out and I couldn't even get a hold of it.

Q. This was about a day or so after?
A. Yes, maybe a couple of days after.

Q. You don't know whether you were shaky before that or not.
A. No, I don't know.

Q. I wonder, one last thing. While you were lying on the cab of that truck, must have been a pretty rough moment there, wonder what sort of thoughts were going through your head then, can you remember?
A. Well, you know one time I saw a tree had blown down on a car and you know it crushed the car in a way, but it didn't go all the way. If that roof was going to come off and land on that truck, and I had my choice of being in here or in that truck, I'd rather be on the floor of that truck. That roof won't come down any further, I don't think than the top of the dashboard. But I don't think I thought of that especially.

Q. You were wishing you had the old car with the narrow windshield?
A. But, I did look up and think, "Oh, how big that windshield is!"

Q. I was wondering if you could remember what thoughts you were having at the moment that you were lying on the floor and the car was bumping up and down and tippin' back.
A. I just wondered if the truck was going to go back over again.

Q. You were remembering these other things then?
A. No.

Q. Worry you?
A. No, I don't think I said a prayer (laugh). It might be a sin not to but I was hanging on.

Q. Far as you know not thinking about anything.
A. That's right, it was just like a drama now that I think of it. You're doing the best you can for yourself.

Q. None of this "life passing in review" like some drama people tell about?
A. Not a bit.

Q. And when it was all over, you thought first of your wife and then your daughter?
A. And then I went up to see my father.

Q. Actually, first of yourself.
A. That's right, first of myself, that's right (laugh).

Q. No, that's perfectly understandable.
A. Yeah.

Q. And all this time you never noticed your heart beating faster or your stomach tightening up or your mouth – getting dry?
A. No.

Q. Have you been sweating any more lately?
A. No more than usual – yeah – at night in bed.

Q. At night in bed?
A. I got awfully warm in bed. I wake up from a… at home we have a, it's a nice cool place where we live. We have a blanket and sheet on the bed. It's been very comfortable when I go to bed, I know it doesn't get any warmer in the room but I woke up awfully warm; and I can't go back to sleep.

Q. You say that you think the disaster work should have had some communication in here for you, I wonder if there is anything else you can think of that could have been done in the way of making disaster work more efficient.
A. Communication, that's about – the thing. After all, I sat on my fat backside here and did nothin' but look after myself. I could have gone up the street and helped some people. If I had realized or could have been informed, even within a half an hour, of the extent of the seriousness of the storm. See, even within half an hour I still thought this was just something that had happened to me.[xxvii] This was my area, here.

In the evolution of the disaster syndrome in this man, who was uninjured or only slightly injured (he does not mention any injury), one can see the syndrome from the inside. Although it becomes apparent, from an analysis of his text, that virtually everything in sight around him was severely damaged or completely destroyed, he thought for half an hour that only he and his property had been damaged (thus blocking out awareness of the community's destruction). At the same time that he said he thought consciously the storm had affected him alone, however, he was expressing. concern over the welfare of his family in other parts of the city. It was half an hour before he was capable of recognizing consciously that this was a community disaster. Days afterward he is dreaming that nothing has happened to anyone but him, and that his teacher (mother) is going to come to his rescue promptly; at the same time, in the dream he is a passive, helpless little boy again.

His altruistic identification with other victims makes him come close to bursting into tears. His behavior seems inexplicable to himself. And he admits that other people thought he was "stunned." His memory for personal experience during impact is very clear, but as far as the record goes, there is partial amnesia for the first half hour after impact with the exception of recollections of trying to locate his family. He did not make any effort to engage in rescue or first aid. During that time, apparently he was quite literally unaware of the extent of the disaster, in spite of the fact that buildings were visibly in ruins and one of them on fire all around him; and he was puttering about aimlessly trying to sweep up twenty-three truckloads of debris with a broom.

2. The Counter-Disaster Syndrome

Persons in the community outside of the impact area at the time of impact, but with close emotional ties to persons and places struck, suffered from a complementary behavioral and emotional syndrome. If the essential behavioral characteristic in the first phase of the disaster syndrome was passivity and (after a first awareness of the extent of damage) ignoring of community trauma, and the responsible mechanism denial and regression, the essential characteristic in the counter-disaster syndrome is over-conscientiousness and hyper-activity, and the responsible mechanism is a defense against feelings of guilt. Fewer observations have been made on the counter-disaster syndrome, partly because many of those who are in a position to describe it are suffering from it and hence are rather defensive in the presence of interviewers or readers.

In its initial stages, the counter-disaster syndrome seems to be characterized by extremely vigorous activity oriented toward rescue, first aid, the making of a contribution of some kind. Certainly there is nothing pathological in activity aimed at helping victim of a disaster; the quasi-pathological quality appears when this activity is, despite the enthusiasm of the helper, relatively low in efficiency and is unduly "panicky," It is interesting that at Worcester, the only two groups whom I have seen described as being "panicky" after impact were absent parents and other relatives returning to the impact area and finding that it had been devastated in their absence, and hospital personnel. "Panic" is not technically the proper word to use: "hyper-active and less rational than normal" might describe it better. Trained personnel, such as firemen and police and local relief personnel, are affected, as well as untrained volunteers.

The first stage, which prevails chiefly during the period of rescue, evacuation and emergency medical care, is characterized by extreme anxiety, with a profusion of autonomic symptoms: tachycardia, shortness of breath, sweating, muscular cramps, etc. The rescue worker is likely to over-exert himself physically, sometimes to the point of collapse; perhaps some of the exhaustion should be ascribed to the large quantity of energy consumed in maintaining internal tension. Far from being passive and dependent, the sufferer is likely to prefer working on his own, or to take a position of responsibility. He does a great deal of work, and its value is considerable, but it is apt to be hastily done and it may require checking and possibly re-doing by less emotionally involved personnel. An example of this in Worcester was the suturing of contaminated wounds by doctors on the night of impact. There are a few descriptions of the counter-disaster syndrome in the interviews. One observer, a psychiatrist, commented on a rescue worker who got into the Great Brook Valley very early:

"... He happened to be driving past and got out and went into the wrecked area that I'm talking about. It was the worst area in terms of destruction, I don't know about the wind, how fast the wind was blowing or where the storm was worst, but this was the worst place. I think they eventually took about fifty dead people out. I didn't think there were that many. I had no idea. But he was overwhelmed by what he saw, as I interpret his behavior, and he just ran around over the tops of the rooms, just running around and I don't know what he thought he was doing, that is, I don't know what he did. I'm sure he thought he was doing the best he could but he didn't do anything, at least at that time..."

There were also various descriptions of the behavior of returning parents: men running up the hill to their homes, and virtually collapsing with exhaustion when they got home; a hysterical woman screaming when a neighbor tried to comfort her; a pastor who spoke of his feeling of "guilt and frustration" at having been away from home when he was needed. To a minor extent, even trained crews of firemen seem to have suffered from this syndrome:

"... it was very hot, we had rubber goods on anyway, you know all our rubber goods, and it made it awfully hot; and the way we were running around there like mad, trying to do what we could... Sweating, oh yes. We were exhausted. We were exhausted, that's what it was... Exhaustion was combine on, because we were rushing, trying to do everything we could, it was really exhaustion that was coming on, see. And I myself was panic-stricken. I was panic-stricken all the time. I didn't know what had happened to my family. I believe the captain felt the same way; now, I didn't speak to the other fellows about it, but they probably felt the same way...

(The captain) was excited. He really was because he... But he was working like mad, and directing us here and ordering us there, you know, and he was carrying out his duty and in the meantime you could see that worried look on his face..."

It should be emphasized that it is the reaction to concern for family, friends, and community that is referred to as the counter-disaster syndrome, not the concern itself. Certainly it is not implied that a sense of responsibility for the welfare of others is undesirable.

Within a few hours, the impact area was swarming with people who had come in from the community outside to help; hospitals had long lines of blood donors, and finally had to turn people away; people called up the Red Cross, Civil Defense, and other agencies, and offered their services. Much of this urge to help simply resulted in obstructing the efforts of trained personnel, or required the detailing of such personnel essentially to combat the counter-disaster syndrome in the community. Hospital corridors and entrances had to be guarded; road blocks had to be established and traffic police assigned; the National Guard had to be assigned to keep out "sight-seers" (many of whom undoubtedly were people anxious to help in some way).

In succeeding days, the continued pressure of the need to demonstrate conclusively the adequacy of one's charity and conscientiousness contributed directly (in my opinion, at least) to the peculiar inter-agency squabbles. These squabbles took two forms: squabbles over the justification for "red tape" (e.g., in Red Cross registration and inquiry practices); and squabbles over the "possession of the disaster" (to use John Powell's happy phrase). The former type of squabble tended to divide local agencies and popular opinion from regional, state, or national offices: thus the National Red Cross was accused of "seeking publicity," was bitterly criticized for trying to register evacuees, for inquiring into financial status of those being assisted, and even for expecting candidates for assistance to apply.[xxviii] And Federal Civil Defense, who requested a housing survey before ordering five hundred trailers sent in from Missouri, were similarly attacked for stalling, heartlessness, etc. One suspects that the relatively unemotional, uninvolved, professional competence of the outsider simply aggravates guilt feelings of the local action- people, who see someone who obviously can't be blamed for what happened, coming in to help and advise (thereby implying the local agencies aren't able to cope with it themselves). Local personnel tend to feel threatened by outside experts, anticipating (not too rationally) criticism of their efforts, and loss of opportunity to make good before the world. Squabbles, however, between local agencies - e.g., between the Red Cross and Civil Defense over welfare administration – also grow out of this competition for the privilege of giving help. In situations where supply is so ample that everyone who wants to help can be given a slice of the material to distribute, it does not matter, but the consequences of competition for short supply could be a disastrous waste of manpower, time, or materials.[xxix]

The only two cases of severe mental breakdown as a result of the tornado (in my records, at least) occurred among people who had shown the counter-disaster syndrome. Traumatic hysterical symptoms (e.g., phobias for thunderstorms) were common among victims from the impact area, but several observers testified that some notoriously unstable personalities seemed to be undisturbed. This is possibly because, in the disaster-syndrome, there is minimal conflict, minimal guilt, simply regression and later restitution. The core of the counter-disaster syndrome, however, is guilt, and this means conflict, and this means the possibility of conflict-induced mental breakdown. The first case was a man who was out of the impact area during the tornado:

A. "… it seemed that everybody was looking for their families. Like this fellow I know that lived up near us at the project, when I was taking my wife and kids out I met him. So, he said, 'How's my wife?' And I said, 'Well, I don't actually know how she is; your mother-in-law is all right,' so it was an assumption that they were both all right. So, he runs like hell up to the project. Well, his wife called up my wife; but he's in Memorial Hospital with a nervous breakdown (laughing). She says, 'The son of a bitch, I went through the whole tornado, he wasn't there, I'm up here with three kids; and he's in the hospital, with a nervous breakdown!'…"

The other case was a woman hospitalized five or six days after impact. The year before impact, her house had been hit by lightning. Her husband had been very active during the tornado rescue period; she herself was ashamed of her timidity.

A. "… And so, as I say, there's nothing I really did. I mean, the whole business - it's just… I think the fact that I am here just because I've been sickly. He was out on the, on the road heading Red Cross units, gathering the ambulances and getting tractors to start and clear the road and all this business and I was all alone with my children an' it… an' all the soldiers coming and going, the policemen and things going around, you know, I couldn't sleep, I couldn't sleep till he got in the house, an' he didn't get in the house until three or four o'clock in the morning. So for five nights neither of us hardly slept a wink. And then, to top it all, I thought I would get out of the area and take my car and go down to Framingham, to Jordan's; it was getting the following week and this week is Father's day so we have a present in mind down there that my daughter and l were going to buy. So we got into the car and managed to get out through the area…you know you have to have a pass to get in and out… and so we got out and got down on to the turnpike and started down a short-ways for… All of a sudden, a car, we have a brand new 1953 Ford, that somehow started going clank, clank, clank, like that, and then it slowed down and wouldn't drive very well. I said to my little daughter, 'Joy, something's wrong with this car' and she said, 'Oh, Mummy, it's just because you bought a Ford.' So I, er, then all of a sudden she said, 'Mummy, people are looking at this car' and so I thought I'd better stop. We stopped. We drew up into what we thought was off the road into a by road, know? Turned out to be a lady's yard and was I glad afterwards because we got out of the car, I shut the motor off and it was still going. It was smoldering and burning like – it wasn't burning; but it was smoldering. I am telling you, I was so upset that I was almost in tears, I guess, since I went through the tornado. I dashed up to the lady's house, we practically banged the windows out in trying to get in the house, and I said, 'Lady, will you send for the Fire Department, my car is on fire.' So she says, 'Oh, calm down, calm down' and so I tried to; but anyway, she sent for the Fire Department and they came out and stood their distance and watched the car. In the meantime, the State Police came up – we had quite an experience – the State Police came up and I said, 'Did that lady send for you?' thinking perhaps she had called the State Police instead of the Fire Department, an' he says, 'No, a truck driver down there told me a car was on fire.' So he came up and he started, he looked at the hood and he said, 'Lady, I just think you're out of water.' I said, 'Oh, no, it can't be; it's a brand new car and I always get water every time I get gas. It can't be.' Well, he waited for the Fire Department and they came up and he said, 'I think she's out of water.' They unscrewed the thing, which was so hot it practically burned their fingers doing it, you know, and that was all that was wrong with it. They put in five gallons of water. So when they finally fixed it up they said; 'I think you can go on for a while, to Framingham if you want to.' But I had no desire to go to Framingham, I wanted to get back home as fast as I could. So I got home and I just left the car plunk in the driveway an' I wouldn't touch it again, so we didn't have too much to eat for supper that night cause I didn't have any (laughing) – I – in fact, I, I didn't even get supper that night because my husband came home and I told him all about it, yet I was too scared to tell him but I told him, and so he started scolding me for not watching the dial or something that it shouldn't get hot but I had really felt that it wasn't my fault, you know, and I felt a little bad about it, so I started to cry; and I cried and I cried. and then, all of a sudden everything went black and I landed here. (laugh and sniffle) I mean they…they called, they called the doctor; I mean all I could remember after that was that he lifted me up and laid me on the couch, I can remember them laying me on the couch. Then they said I'd better come here and rest for a few days. So, as they said, they said it was too close. So they said it was once too much. They said it was once too much. They came to find out it wasn't my fault at all, it was the, er, on the thousand-mile check- up, on the car, they hadn't bolted down some part of it, or whatever it was. My husband was awful mad, he said, 'I won't settle for anything less than a new engine in that car.' But they said the car wasn't damaged, I thought so but it wasn't harmed. But that was just the climax which landed me up here…"

Q. Well, I think your story is most helpful to us because you're a person who was not hit directly by the tornado but it shows the emotional impact even when there wasn't a physical impact on people, you see.

A. That's what the doctor said. I just talked to him today. And he said, 'I was telling him 'cause I feel like such a sissy, that the people that were hit just a little bit were lucky because they were taken to the hospital right off and that emotional upset was subdued whereas mine has been building up for five or six days.'

Q. Until something just had to come along and in a way in which you felt guilty about, just some little thing, and it was just too much for you to take. Under ordinary circumstances you could ride it.

A. That's right, under ordinary circumstances I would…it would have just slid off of me, and I'd have got a little mad about it, most likely; but I didn't… I took it differently.

Q. How are you feeling now, Mrs. X?

A. Oh, I'm feeling fine now. Don't know exactly when I'm going to go home, but I hope it won't be too long. They took some punctures out of my spine and took some pictures of my head to see why I…why I blacked out, I guess, and if there was anything wrong, but there wasn't, it was just an emotional upset.

A. …just a year before that we'd had a tree that well, lightning it hit the house, the aerial on the house, and went down through the drain pipe and out through the septic tank and hit the – blew a big hole in the back yard, and hit a tree. I got excited when I knew the storm was coming, so I made my nephew go up and get her and bring her home, much against his will; he said he had a lot of…

Q. You knew a storm was coming?

A. I thought it was a thunder storm because the radio said it was going to be a bad thunderstorm. So, and you could see it was so black over there. So I had him go up and get her, and he was so provoked, he said he had a lot of letters to write and get out in the mail before the storm broke, but I made him go, and I'll bet that he's glad today. So that we were all at home and all together when it happened. And we… sometimes I wonder if we'd been right plunk in the middle of it, right in that whirling section of it - we were right on the edge of the whirl, whether we'd have ever made the cellar soon enough. I wonder that because we were upstairs and … er … the living room… something crashed through the living room and we were standing near the dining room and the trees went like this and broke in two right in the living… in the dining room. So then my husband said… before that, I kept saying to my husband, 'Let's go down cellar, let's go down cellar,' 'cause I didn't like the looks of it. And so when it happened, my husband said, 'C'mon let's go to the cellar as fast as we can.' But lots of people they say, you know…

Q. So you did get down to the cellar?

A. We did get down to the cellar, yes. A little more went on while we were down in the cellar, because all this black…

Q. What were you saying to one another while you were down in the cellar? Do you recall?

A. It all happened so quickly, I mean it was all over within five or six minutes. I mean we just got down cellar and all this black stuff came at you, and he kept sayin' 'Let's get in this other corner of the cellar, let's get in this other corner of the cellar.' He was tryin' to think in his mind which corner would be the safest. We were mostly caught in the middle anyway. But then it was all over with before you could…

Q. Did you talk about what you thought happened to you all?

A. No. You didn't even have time. The only time we talked at all was before we went down to the cellar. We were standing in the dining room before the trees went down, and my husband knows I'm timid and so I tried to be brave and started talking about somethin'…In fact the funny part of it was before it started, I was sitting in the living room and I thought, 'Well, I'll just sit still and try to behave myself.' I didn't want to upset the children too much so I opened the evening paper and here it was Flint, Michigan you know an' all the… (laughing). But even then that didn't bother me, because I didn't realize what was happening. Well, then it started getting a little bit – oh no, then the hailstones came. Ooh! Huge hail stones big as golf balls – actually as big as golf balls, I never saw any so big in my life, in fact my son opened the back door and brought one in. So that made me pop up from the living room. So we stood there in the dining room watching it, well watching the hail stones. My husband says, 'Y' know this is a funny storm.' He watches weather all the time, because he has an airplane and he flies a lot, so he watches weather all the time, he has a special radio in his bedroom that indicates weather, but that still didn't tell him that it was going to be this kind of a storm, or at least he's not enough educated in the weather to know. But he says 'This is a funny storm, you know it's coming from the north and the wind is coming from the south.' That's about all he had time to say cause the trees started going like that and broke off like sticks. So we went down to the cellar. Then when we came up it was all over. Well, it had passed us, it must have still been going on beyond us, when you stop and think, because we saw that house go down…

    The desire to be of assistance and to receive recognition for having been a competent actor in the emergency is a powerful one, and if it is not satisfied, disturbed behavior can result (presumably, in presensitized persons particularly) as resentment and guilt build up. Survivors from outside the impact area need to "get into the act" and interference with their efforts to play a satisfying relief role may precipitate emotional conflicts which reduce efficiency in rescue and relief operations.

3. The Length of the Isolation Period.

It may be contended that one of the most crucial factors governing the incidence of casualties and property damage is the length of the isolation period. The logic behind this statement is as follows. If two tornadoes (or any other impact agent) strike two inhabited areas, occupied by the same number and sorts of people and structures, but if a given quantity of protective personnel and equipment moves into the impact area after half an hour in the first case, while they move in within five minutes in the second: there will inevitably be a larger number of casualties and more property damage in the first case than in the second. In symbolic form,

$$D = f(I \cdot P)$$

where (D) is the quantity of damage in some category of phenomena (e.g., number of deaths), (I) is the length of the isolation period in hours, (P) is the pre-impact quantity of the phenomenon. This relationship should be valid because within the impact area, during the isolation period after a sudden impact, a continuous process of secondary impact will increase damage: fires, exposure, sepsis in wounds, shock, and the continuation of lethal processes (like bleeding, asphyxiation, etc.) set in motion by the primary impact. The chief function of the rescue force is to terminate secondary impact before it increases casualties and damage above the amount left by primary impact. Anything which lengthens the isolation period for an impact area, or a part of an impact area, will thereby increase the incidence of injury or damage. An added fifteen minutes of isolation in Worcester would certainly have substantially increased both property damage and casualty lists; one recalls that the fire engines were "barely" in time to prevent a major conflagration in the Greenhill-Burncoat areas (the three-house fire had almost passed beyond the reach of hose lines from the few still-functioning hydrants). If, instead of a warm June evening, it had been windy, subzero weather and night-time, exposure would have been an important factor, and time here would have been of critical importance. One would venture to predict that, in the absence of aid from outside the impact area, damage would increase somewhat as follows:

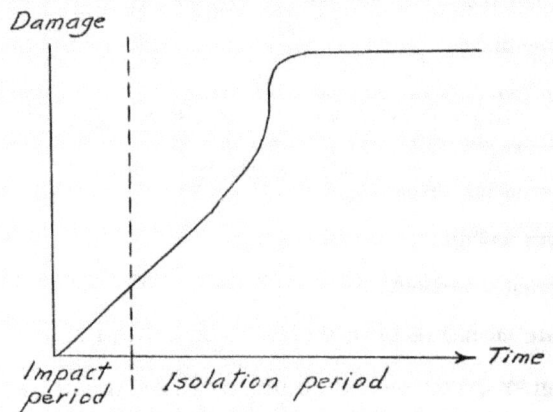

The reason for the acceleration-point in the isolation period is that certain types of secondary impact (bleeding, asphyxiation, lethal wound shock, and fires) should, if unimpeded, combine to produce a second wave of injuries several minutes after impact. The reason for the leveling of the curve is that the dazed state will eventually "wear-off" by itself, and after this point an increasing number of victims will be able to care for themselves, while many of the most serious consequences of secondary impact will already have run their course.

Various factors affect the length of the isolation period at any particular point in an impact area. One of these is the distance from that point to the edge of the impact area – in other words, the larger an impact area, the longer the average isolation period for all points in the impact area. Another factor, of course, is the extent to which the impact has interfered with communication and transportation. A third factor is the distance from the edge of the impact area to the protective units. A fourth is the quickness with which the protective units are notified; a fifth, the quickness with which they mobilize; a sixth, the completeness of the reconnaissance and inventorying by the protective units, both before and after reaching the impact area. The list can obviously be extended to include a great many other actors which affect the length of time which passes before a given unit of impact area space is reached by outside aid.

It is evident also, that many other things, having no relation to the length of the isolation period, also affect the amount of damage: the nature of the impact agent, the kind of structures in the impact area, composition of the population in the impact area, the efficiency of protective personnel and equipment, and the quantity of protective personnel and equipment brought to bear, at any given point, are particularly important. I have emphasized the time factor in the relationship between the protective agencies and the secondary impact: here there is, as it were, a race between rescue agencies and secondary impact agents to reach potential targets.

4. The "Cornucopia Theory"

The rescue and rehabilitation operations at Worcester were considered to have been relatively successful. The isolation period was short, rescue and evacuation was fast, medical care was quickly made available to all victims requiring it, and the rehabilitation procedures of every kind were furnished with lavishness. While specific instances of inefficiency were pointed out, very few instances of failure could be. The principles of "wave supply" and "from heaven", pointed out by Rosow, meant that even when efficiency was low, the sheer mass of services and materiel were able to satisfy needs as they came up. Comparison might be made with military firepower: a good marksman with a finely-tooled and sighted rifle may be more efficient, but a machine gun gets better results in holding down an enemy position – provided there is more ammunition available than is needed to account for each enemy soldier. In disaster operations, when materiel and personnel are pouring out of a cornucopia, deluging the impact area, the results in rescue and rehabilitation are almost inevitably impressive.

This is what happened at Worcester. The impact area was blanketed with protective agencies: hundreds of police, firemen, National Guards, public works people, CD volunteers, and miscellaneous helpers invaded it during the rescue period; hospitals had more blood donors than they could handle; the Red Cross mobilized hundreds of nurses; equipment and supplies of all kinds were funneled into Worcester from all over the Northeast, and four hundred twenty-five trailers came from Missouri. While the results of this sort of provision are so, good, that post-mortem studies have little to criticize except relatively minor matters and little to recommend except more efficient utilization of what was already available, they take for granted the fact that the cornucopia principle's successful application at Worcester depended on the fortunate (and not at all inevitable) co-existence of two conditions: a complete lack of damage to Worcester's own protective agencies, and to those of any other source of regional aid; and the absence of any competition from anywhere nearer than Ohio for emergency supplies and personnel.

A glance at the map of the city, showing the path of the tornado and the location of protective agencies, will show that this impact could, however, have wiped out or severely crippled, most of the police stations (including ambulances), fire stations, hospitals, Red Cross and CD headquarters, and government centers, if it had passed through Worcester on a different course. If it had taken such a course, also, the number of primary-impact casualties (to say nothing of the results of secondary impact) would probably have been much greater. Furthermore, if the tornado had proceeded another thirty miles, into the Boston area, considerable quantities of supplies, personnel, and equipment which in reality found their way to Worcester, would probably have stayed in (or gone in) to Boston.

The cornucopia theory thus rests on the two assumptions that any given disaster will not destroy the cornucopia itself and that any given disaster or combination of disasters will be unable to exhaust the cornucopia before adequate relief and rehabilitation can be provided.

I have the feeling that the theory is widely held if rarely formally stated. In all probability, these assumptions are valid for most natural disasters (fires, floods, earthquakes, tornadoes, hurricanes, tidal waves, epidemics, etc.).

It is a question, however, whether the assumption does apply to disasters which might be produced by atomic or hydrogen bomb explosions.

In such events, it might well be that the cornucopia would be itself largely smashed and its supplies exhausted long before the secondary impact was under control.

Now pointing out the potential inadequacy of the cornucopia does not imply that there is anything wrong with having a cornucopia. The questions which I should like to raise, however, are:

(1) Does the faith in the cornucopia, as experienced in natural disasters, produce a tendency to think in terms of repair rather than prevention?
(2) Does the faith in the cornucopia tend to produce organizations which are better adapted to excess supply than to inadequate supply?

In other words, there is a basic question whether the type of organization and planning which gets results, where there is more than enough of personnel and supply will be most effective when everything is short.

I don't have ready answers to these questions. But it is striking, in the case of the Worcester tornado – a severe natural disaster, but minor in comparison with what a military disaster would be – that for most rescue and rehabilitation functions, there were several responsible agencies with overlapping jurisdictions, and usually more than enough personnel and supplies to go around. Indeed, many people were kept busy simply acting as organizational traffic policemen, to keep people and agencies off each other's toes. Everything seemed to move in an atmosphere of: "There's plenty to go around, and if I run short, I'll call Joe on the phone and he'll send some over." The cornucopia nourishes a sort of autonomy and duplication of organizations in the midst of plenty. If Red Cross and Civil Defense dispute over the supervision of welfare activities, the solution can wait for thirty-six hours, and finally both can be given some responsibility in this area.

This is fine when supply exceeds necessity.

But if there weren't enough food or clothing, or shelter in the area, and two autonomous agencies squabbled over what little there was, not only would the utilization of that little be inefficient, but the personnel of one of the organizations would be wasting their time while they could be doing something else.

Furthermore (referring to question 1 above), a profusion of rescue and relief agencies seems to be conducive to an atmosphere of "waiting until it happens" before doing anything about it. Reams have been written about the behavior of organizations after the tornado; but only passing attention has been paid to analyzing the factors which allowed a tornado to march for an hour through central Massachusetts without any part of the general population ever being warned to take cover. A combination of radio announcements and telephone calls from central exchange points could have had perhaps 90 per cent of the population in cellars within minutes. Such a warning would have saved more lives than any conceivable improvement of procedures in the rescue and rehabilitation phases.

In other words, I wonder whether we ought not to put more stress on "stop it from happening," even while we keep "repair and replace" procedures at the highest possible peak of effectiveness. And this would apply with even greater force to anticipations of military impacts, following which the "repair and replace" cornucopia may not be there anymore.

October 1954

BIBLIOGRAPHY (Original)

American National Red Cross. Massachusetts - New Hampshire Tornado of June 1953: Final Report. Worcester: Disaster Headquarters, n. d. /1953/. Mimeographed.

American Red Cross. Worcester Tornado News. Worcester: Disaster Headquarters, 1953. 9 bulletins.

Bakst, H. J., Berg, R. L., Foster, F. D., and Raker, J. W. The Worcester County Tornado: A Medical Study of the Disaster. Mimeographed, 1954. 81 pp. This study was sponsored by the Committee on Disaster Studies.

Bowman, H. L. Physical Damage from Central Massachusetts. Tornado of June 9, 1953. Washington: Unite States Atomic Energy Commission, 1953. Mimeographed, photographs. 3pp. text, 18 plates.

Brodsky, C. M., Muldoon, J. F., and Herzfeld, R. F. An Exploratory Study of the Role of Catholic Church Organizations in Disaster. Washington: Catholic University of America, n. d. /^1953/. Mimeographed. 73 pp. This study was sponsored by the Medical Research and Development Branch, Office of the Surgeon General, Department of the Army, at the suggestion of the Committee on Disaster Studies.

Flora, Snowden D. Tornadoes of the United States. Norman: University of Oklahoma Press, 1953.

Harold, R. P. "Downtown Worcester Due for Transformation." American City. 67 (1952): 101.

Knight, Richard C. "Observation of Worcester, Massachusetts, Tornado Disaster." Memorandum to Rescue Division, Federal Civil Defense Administration. 22 June 1953. 2 pp.

Cranach, Wilmer L. "The Role of the Protestant Churches in the Central Massachusetts Tornado, June 9, 1953." Memorandum to Welfare Division, Federal Civil Defense Administration. 21 June 1953. 5 pp.

Landstreet, Barent F. "Field Trip - Central Massachusetts Tornado of June 9 5 1953." Memorandum to Evacuation Planning Branch, Federal Civil Defense Administration. 1 July 1953. 15 pp.
This survey was done in cooperation with the Committee on Disaster Studies.

Morris, F. D. "City That Didn't Cry Uncle." Colliers, 125(1950): 34-35.

Powell, John W. "Investigation of the Worcester Tornado, June, 1953: Preliminary Narrative and Impressions." Memorandum to Committee on Disaster Studies, National Research Council. 20 July 1953. 16 pp.

_____, and Rayner, Jeannette. "Progress Notes: Disaster Investigation." Contract report, Chemical Corps Medical Laboratory, Army Chemical Center, 1952.

_____, Rayner, Jeannette, and Finesinger, Jacob E. "Responses to Disaster in American Cultural Groups." In Symposium on Stress (Washington: Army Medical Service Graduate School, 1953.

_____, See "Interviews and Field Notes."

Rayner, Jeannette F. See "Interviews and Field Notes."

Rosow, Irving. Communications in the Worcester Emergency. MS, 1954 (draft of extensive report to Committee on Disaster Studies).

Schultz, P. L. "Tornado Hits Three Worcester Housing Projects." Journal of Housing, 10 (1953): 223.

_____, "Mobilizing for Disaster." Journal of Housing, 10 (1953): 264.

Wallace, Anthony F. C. "Memorandum on Worcester Disaster Study." Memorandum for Committee on Disaster Studies, National Research Council. July, 1953. 6 pp.

_____, Human Behavior in Extreme Situations: A Survey of the Literature and Suggestions for Further Research. Washington: Committee on Disaster Studies, 1953. Mimeographed.

_____, See "Interviews and Field Notes."

Worcester City Directory, 1953. Boston: R. L. Polk & Co., 1953.

"Worcester Plants Pick Themselves Up." Business Week, 20 June 1953, pp. 30-31.

"Worcester Tornado." American City, 68 (1953).

Worcester Telegram and Evening Gazette, 10 June - 18 June 1953.

WTAG (radio station), typescript of tape recordings of "Tornado 1" and "Tornado 2" (special information broadcasts on the tornado)

INTERVIEWS AND FIELD NOTES (ORIGINAL)

During the preparation of the report, several collections of transcripts of original interviews and questionnaire responses were made available to the writer.

Maryland, University of, Disaster Research Project (sponsored and given staff assistance by the Committee on Disaster Studies).

Personnel and friends of the Disaster Research Project, University of Maryland, headed by Dr. John Powell, made a number of interviews in Worcester. These interviewers (Dr. John Powell, Dr. Enoch Callaway, and Mrs. Edna Barrabee) had all had considerable experience with "in-depth" interviewing in clinical psychiatric situations. These interviews were relatively unstructured in that the interviewers allowed the respondent to develop his narrative and associations freely, but the interviewer guided the interviews as a whole by bringing up and probing for material on a series of topics about which information was required. The following interviews were available to the writer:

Interviewed by John W. Powell: Five firemen and one individual from the fringe impact area.

Interviewed by Enoch Callaway: Three individuals, all victims of the tornado, one a physician.

Interviewed by Mrs. Edna Barrabee: Twenty-five individuals, twenty-three of whom were hospitalized Victims; one individual from the community aid area, and one individual related to a victim.

The team sent into the field by the Committee on Disaster Studies made several separate surveys. Available to the writer, from this group of studies, were transcripts of recorded interviews by Miss Jeannette Rayner and of course his own interviews:

Interviewed by Jeannette Rayner: Eighteen individuals, three of whom were hospitalized victims; the others, persons in official medical positions.

Interviewed by Anthony Wallace: Six individuals, all occupying community leadership positions.

NATIONAL ACADEMY OF SCIENCES – NATIONAL RESEARCH COUNCIL

The National Academy of Sciences-National Research Council is a private, nonprofit organization of scientists, dedicated to the furtherance of science and to its use for the general welfare.

The Academy itself was established in 1863 under a Congressional charter signed by President Lincoln. Empowered to provide for all activities appropriate to academies of science, it was also required by its charter to act as an adviser to the Federal Government in scientific matters. This provision accounts for the close ties that have always existed between the Academy and the Government, although the Academy is not a governmental agency.

The National Research Council was established by the Academy in 1916, at the request of President Wilson, to enable scientists generally to associate their efforts with those of the limited membership of the Academy in service to the nation, to society, and to science at home and abroad. Members of the National Research Council receive their appointments from the President of the Academy. They include representatives nominated by the major scientific and technical societies, representatives of the Federal Government designated by the President of the United States, and a number of members-at-large. In addition, several thousand scientists and engineers take part in the activities of the Research Council through membership on Its various boards and committees.

Receiving funds from both public and private sources, by contributions, grant, or contract, the Academy and its Research Council thus work to stimulate research and its applications, to survey the broad possibilities of science, to promote effective utilization of the scientific and technical resources of the country, to serve the Government, and to further the general interests of science.

– End of Original Document –

ADDITIONAL RESOURCES (NEW)

John Brooks, *Five-Ten on a Sticky June Day,* New Yorker Magazine, May 28, 1955

Kyle Beatty, *Twist and Shout,* Risk Management Solutions, Inc., Newark, Calif., in Global Reinsurance., Nov. 2001

Brown University Sciences Library, Providence, R.I.

The late Rita Canney, Rutland, Mass. Climatological Data, National Summary, June, 1953

Climatological Data, New England, June, 1953

Steve Corfidi, *A Brief History of the Storm Prediction Center,* NOAA, Norman, Okla.

Mr. & Mrs. Gilbert S. Davis, Worcester, Mass.

Davis Press, *Tornado!* Worcester, Mass., 1953

John Drury, Druid Photography, Providence, R.I.

Raymond E. Falconer & Vincent J. Schaefer, *Observations of the Formative Stages of the Worcester Massachusetts Tornado,* Final Report, ONR Project, G.E. Research Laboratory, Schenectady, N.Y., Dec.1953

Thomas P. Grazulis, *Significant Tornadoes: 1680 – 1991*, Environmental Films, St. Johnsbury, Vt.

John Hales, Storm Prediction Center, Norman, Okla.

Harvard Forest, Petersham, Mass. Holden Historical Society, Holden, Mass.

New England Fire Insurance Rating Association, *Tornado - Central Massachusetts Catastrophe No. 42,* Boston, Mass., 1953

John O'Toole, *84 Minutes, 94 Lives,* Databooks, Worcester, Mass., 1993

Samuel Penn, Charles Pierce & James McGuire, *The Squall Line and Massachusetts Tornadoes of June 9, 1953,* Bulletin of the American Meteorological Society, Boston, Mass., March, 1955

John W. Raker, M.D., *Lecture at the Walter Reed Army Institute of Research,* Walter Reed Army Medical Center, Washington, D.C., Jan.1955

Howard Smith, Portland, Maine

Disaster Research Center Library, University of Delaware, Newark, Del.

Anthony F. C. Wallace, *Tornado in Worcester,* National Research Council, Washington, D.C., 1956

Worcester Public Library, Worcester, Mass.

Worcester Telegram & Evening Gazette, Worcester, Mass.

World Book Encyclopedia, 1954 Annual Supplement

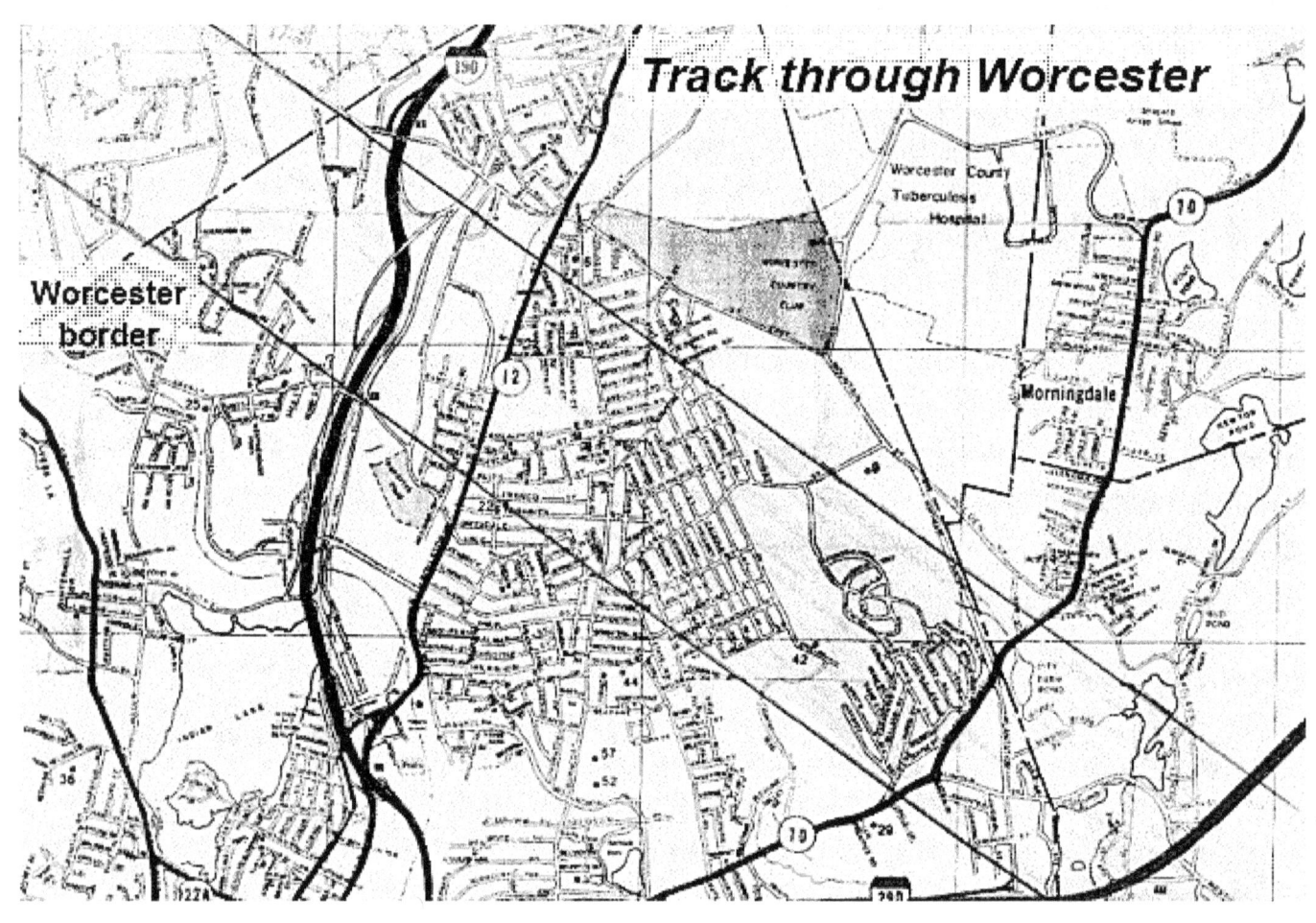

NEW:

SPECIAL MAP: THE PATH OF THE 1953 TORNADO SUPERIMPOSED ON THE MAP OF
WORCESTER, MASSACHUSETTS (PRESENT DAY)

ENDNOTES (NEW)

[i] Throughout the report, the background context is that of the "Cold War" mentality – what effect would a nuclear strike have on the general population. The nearest thing that anyone could equate that event to was a tornado in a populated area.

[ii] My memories of Worcester, while I was growing up, were that of a "shining city", bursting with civic pride. It was remarkably clean and well organized. There seemed to be a sense that great things were happening. Sadly, by the time I had returned from the military (1971), the city had greatly decayed following several major "strokes" of massive plant shutdowns and forced unemployment. With the manufacturing base stripped and an aging middle class the city was rapidly declining. Today the city seems to be stabilized, but it still shows its age. One of the most confusing aspects, however, remains the maze of "Private Streets" that sometimes form a crippling maze, especially where they deteriorate into gravel "drives". I was never able to fully understand how this could be a viable alternative for a major city of the size of Worcester. These private streets/roads were prevalent in the hillside area directly below Uncatena Avenue and adjacent to the Great Brook Valley Area.

[iii] Worcester had a high concentration of members of the Catholic faith, balanced (equally) by numerous, active protestant denominations, some dating back to colonial times. It was a predominantly "A Christian City", and the ethics of the general population reflected charity and "good-will".

[iv] Note the use of terms used to describe the various neighborhoods. Certainly NOT "Politically Correct!"

[v] As of the 2010 Census the city's population was 181,045, making it the second largest city in New England after Boston. Worcester is located approximately 40 miles (64 km) west of Boston, and 38 miles (61 km) east of Springfield.

[vi] According to the City's 2010 Comprehensive Annual Financial Report, the top ten employers in the city are:

#	Employer	# of employees
1	UMass Memorial Health Care	13,764
2	University of Massachusetts Medical School	5,678
3	City of Worcester	5,128
4	Saint Vincent Hospital	2,386
5	Hanover Insurance	1,850
6	Saint-Gobain	1,807
7	Reliant Medical Group	1,801
8	Polar Beverages	1,400
9	College of the Holy Cross	1,107
10	Quinsigamond Community College	900

[vii] Worcester is served by several interstate highways. Interstate 290 connects central Worcester to Interstate 495, I-90 in nearby Auburn, and I-395. I-190 links Worcester to MA 2 and the cities of Fitchburg and Leominster in northern Worcester County. I-90 can also be reached from a new Massachusetts Route 146 connector. Worcester is also served by several smaller Massachusetts state highways. Route 9 links the city to its eastern and western suburbs, Shrewsbury and Leicester. Route 9 runs almost the entire length of the state, connecting Boston and Worcester with Pittsfield, near the New York state border. Route 12 was the primary route north to Leominster and Fitchburg until the completion of I-190. Route 12 also connected Worcester to Webster before I-395 was completed. It still serves as an alternate, local route. Route 146, the Worcester-Providence Turnpike, connects the city with the similar city of Providence, Rhode Island. Route 20 touches the southernmost tip of Worcester near the Massachusetts Turnpike. U.S. 20 is a coast-to-coast route connecting the Atlantic to the Pacific Ocean, and is the longest road in the United States.

Worcester is the headquarters of the Providence and Worcester, a Class II railroad operating throughout much of southern New England. Worcester is also the western terminus of the Framingham/Worcester commuter rail line run by the Massachusetts Bay Transportation Authority. Union Station serves as the hub for commuter railway traffic. Built in 1911, the station has been restored to its original grace and splendor, reopening to full operation in 2000. It also serves as an Amtrak stop, serving the Lake Shore Limited from Boston to Chicago. In October 2008 the MBTA added 5 new trains to the Framingham/Worcester line as part of a plan to add 20 or more trains from Worcester to Boston and also to buy the track from CSX Transportation. Train passengers may also connect to additional services such as the Vermonter line in Springfield.

The Worcester Regional Transit Authority, or WRTA, manages the municipal bus system. Buses operate intracity as well as connect Worcester to surrounding central Massachusetts communities. The WRTA also operates a shuttle bus between member institutions of the Colleges of Worcester Consortium. Worcester is also served by Peter Pan Bus Lines and Greyhound Bus Lines, which operate out of Union Station.

The Worcester Regional Airport, owned and operated by Massport lies at the top of Tatnuck Hill, Worcester's highest point @ 1,000 ft. elevation. The airport consists of one 7,000 ft. (2,100 m) runway and a $15.7 million terminal. JetBlue began daily service from ORH to Fort Lauderdale, and Orlando, Florida beginning in November 2013.

[viii] Worcester is home to several institutes of higher education.
- Assumption College is the fourth oldest Roman Catholic college in New England and was founded in 1904. At 175 acres (0.71 km2), it has the largest campus in Worcester.
- Becker College is a private college with campuses in Worcester and Leicester, Massachusetts. It was founded in Leicester in 1784 as Leicester Academy. The Worcester campus was founded in 1887, and the two campuses merged into Becker College in 1977. Becker's video game design program has consistently been ranked in the top 10 in the U.S. and Canada.
- Clark University was founded in 1887 as the first all-graduate school in the country; it now also educates undergraduates and is noted for its strengths in psychology and geography. Its first president was G. Stanley Hall, the founder of organized psychology as a science and profession, father of the child study movement, and founder of the American Psychological Association. Well-known professors include Albert A. Michelson, who won the first American Nobel Prize in 1902 for his measurement of light. Robert H. Goddard, a pioneering rocket scientist of the space age also studied and taught here, and, in his only visit to the United States, Sigmund Freud delivered his five famous "Clark Lectures" at the University. Clark offers the only program in the country leading to a Ph.D. in Holocaust History and Genocide Studies.
- The Jesuit College of the Holy Cross, was founded in 1843 and is the oldest Roman Catholic college in New England and one of the oldest in the United States. Well-known graduates include Dr. Joseph E. Murray, Nobel laureate in Physiology or Medicine, Billy Collins, former Poet Laureate of the United States, Bob Couse, and Supreme Court Justice Clarence Thomas. In 2013, the College of the Holy Cross was ranked by U.S. News and World Report as the nation's 25th highest-rated liberal arts college.
- The Massachusetts College of Pharmacy and Health Sciences Worcester Campus houses the institution's accelerated programs in Nursing and Doctor of Pharmacy as well as the Master's program in Physician Assistant Studies for post-baccalaureate students.
- Quinsigamond Community College.
- The University of Massachusetts Medical School (1970) is one of the nation's top 50 medical schools. Dr. Craig Mello won the 2006 Nobel Prize for Medicine. The University of Massachusetts Medical School is ranked fourth in primary care education among America's 125 medical schools in the 2006 U.S. News & World Report annual guide "America's Best Graduate Schools".
- Worcester Polytechnic Institute (1865) is an innovative leader in engineering education and partnering with local biotechnology industries. Robert Goddard, the father of modern rocketry, graduated from WPI in 1908 with a Bachelor of Science in physics.
- Worcester State University is a public, 4-year college founded in 1874 as Worcester Normal School.

[ix] It would appear there has been a marked "shift" in voter preference since 1953.

| Voter registration & party enrollment as of August 20, 2014 - Worcester |||
Party	Number of voters	Percentage
Democratic	47,275	44%
Republican	9,310	9%
Unenrolled	50,397	47%
Political Designations	0	0%
Totals	107,686	100%

[x] FIRE DEPARTMENT: (2013 – Latest Statistics)
- Employees   424
- EMS level   BLS
- Facilities and equipment
    - Stations   10
    - Engines   13
    - Trucks   3
    - Tillers   2
    - Platforms   2
    - HAZMAT   1

[xi] The department employed 335 officers and 85 officials (420 total personnel), in the fiscal 2013 budget.

[xii] Since the end of the Cold War, civil defense has fallen into disuse within the United States. Gradually, the focus on nuclear war shifted to an "all-hazards" approach of Comprehensive Emergency Management. Natural disasters and the emergence of new threats such as terrorism have caused attention to be focused away from traditional civil defense and into new forms of civil protection such as emergency management and homeland security.

[xiii] This is an obvious reference to the Cold War hysteria of impending nuclear attack.

[xiv] Today, the American Red Cross is the only non-governmental organization chartered by Congress to provide relief to victims of disaster. Under the Federal Response Plan following a Presidential declaration, the Red Cross and the Federal Emergency Management Agency (FEMA) work cooperatively. The American Red Cross is a signatory to the Federal Response Plan and is obligated as an agent of the federal government to coordinate all mass care response assistance through Emergency Support function (EFS) #6. This means the Red Cross is the primary agent designated to provide mass care relief including food, shelter, supplies, first aid and more. The Red Cross also operates a Disaster Welfare Information System for the purpose of reporting victim status and assisting in family reunification.

[xv] Forecasters at the National Weather Service office in Boston believed that there was a possibility for tornadic activity in the area, but decided not to include it in their forecast for the day in fear that they would cause panic among local citizens. 1953 was the first year that tornado and severe thunderstorm warnings were used, so forecasters compromised and issued the first severe thunderstorm watch in the history of Massachusetts. Because of this, the tornado struck with little to no warning for residents.

[xvi] Fayville is located close to the center of Massachusetts. Fayville is considered a part of Southborough and is located at the intersections of Route 9, Central Street/Oak Hill Road. Fayville is within Worcester County.

[xvii] Apparently, "…no one knew what to do…" when confronted with this situation! No independent thought was given to spread the alarm or to seek help outside the local area. This could only be attributed to a complete lack of understanding of the scope of the impending disaster and the concern for "self" overriding any other action.

[xviii] It never ceases to "amaze" (me in particular), how someone could NOT interpret an approaching tornado for what it was. I can only imagine I was the only person ever to see "The Wizard of OZ" movie!

[xix] Thus was fostered the greatest of "true" urban legends – the brain sucking tornado.

[xx] I find it interesting that the report writer DID NOT consider the following; The Tornado was (only) 8 years after WW II and during the Korean wartime. It should have been considered that many of the men in the initial impact area HAD "disaster training" based on real-time military experience. This "pool" of trained persons probably was of greater benefit to those within the impact area since formal rescue units were hampered by the massive debris and destruction focused in a small area.

[xxi] Why no thought to employ Boy Scouts as "Messengers" or "Runners" escapes me since this is a fundamental method of communication.

[xxii] "The Worcester County Tornado – a Medical Study of the Disaster," Henry K. Bakst, Fred D. Foster and John W. Raker; 1955.

[xxiii] The Worcester Tornado of 1953 created its own "House Trailer" argument that predated the Hurricane Katrina debacle. At the time Hurricane Katrina struck, there were about 14,000 trailers in lots around the country, waiting to be sold; FEMA determined it needed 120,000 to accommodate displaced homeowners and for temporary shelter. It ordered nearly $2.7 billion worth of travel trailers and mobile homes from 60 different companies. FEMA ultimately succeeded in deploying 140,000 trailers up and down the ravaged Gulf Coast. Then it had to start figuring out what to do with them as people began to rebuild their lives and leave them behind.

[xxiv] To put things in perspective: Recovery from the Worcester tornado was (fortunately?) a factor of the storm itself. For example: Hurricanes tend to cause much more destruction than tornadoes because of their size, duration and the variety of ways to damage items. The destructive circular eyewall in hurricanes (that surrounds the calm eye) can be tens of miles across, last hours and damage structures through storm surge, rainfall-caused flooding, as well as wind impacts. Tornadoes, in contrast, tend to be a mile or smaller in diameter, last for minutes over any given point and primarily cause damage from their extreme winds. Beyond their immediate vicinity, tornados have less "secondary damage".

[xxv] After a temporary stay at 1010 Main St, the College officially moved to its current Salisbury Street address. This would not have been possible without the contributions of the local community and that of the Kennedy family. Following the Senator's tour of the crippled campus, local Bishop John Wright united his mutual friends in the Kennedy family and at the schools. It was arranged that a donation of $150,000-the largest the college had received to date-be made to Assumption for the rebuilding effort. This donation was made on July 20, 1953, a little more than a month after the disaster, at Worcester's Sheraton Hotel. The first on-campus commencement was held in the new dining hall in June of 1957.

[xxvi] Post-traumatic stress disorder (PTSD) is a psychiatric condition that can develop following any traumatic, catastrophic life experience. Recognition of this condition increased dramatically following the war in Viet Nam, when many returning U.S. veterans developed disturbing psychological symptoms and impaired functioning. More recently, the 9/11 tragedy, the Asian tsunami, the London bombings, and Hurricane Katrina and its aftermath have left thousands of people at risk for this potentially debilitating condition.

PTSD symptoms can develop weeks or months, or sometimes even years, following a catastrophic event. Along with survivors of natural disasters, wars, and acts of terrorism, people who have been the victims of violent crime or torture often develop symptoms of PTSD.

PTSD symptoms vary among individuals and also vary in severity from mild to disabling. PTSD Symptoms can include one or more of the following:

- "flashbacks" about the traumatic event
- feelings of estrangement or detachment
- nightmares
- sleep disturbances
- impaired functioning
- occupational instability
- memory disturbances
- family discord
- parenting or marital difficulties

Sometimes the manifestations of PTSD wax and wane, with symptom-free intervals occurring between symptomatic episodes.

Anniversaries and reminders of the precipitating event can exacerbate the symptoms. Sometimes PTSD occurs in combination with other emotional disorders or with specific physical symptoms.

PTSD can develop in persons of any age, including children. The diagnosis of PTSD is confirmed when the disturbing symptoms persist for longer than one month. Because individuals differ in their reactions to traumatic events, it is not possible to predict in advance who will develop PTSD following a tragedy. PTSD is relatively common, with 3.6% of U.S. adults estimated as having PTSD in a given year.

Disasters that involve high winds tend to cause a range of effects depending on their severity. After disasters of this type, the impairment of most survivors is moderate. Effects may be prolonged but do not usually meet criteria for a mental disorder. However, in a significant minority of research studies on disasters of wind, at least 25% of the survivors had symptoms of a diagnosable disorder. Symptoms usually peak during the first year, and in the vast majority of cases, survivors get better with time after the disaster. (Norris, F., Friedman, M., Watson, P., Byrne, C., Diaz, E., & Kaniasty, K. (2002). 60,000 disaster victims speak, Part I: An empirical review of the empirical literature, 1981 - 2001. Psychiatry, 65, 207-239. AND Norris, F., Friedman, M., & Watson, P. (2002). 60,000 disaster victims speak, Part II: Summary and implications of the disaster mental health research. Psychiatry, 65, 240-260.)

[xxvii] There is no doubt in my mind that a disaster of this proportion would be well "communicated" to the general public by a plethora of agencies/news outlets and "social media" outside the local area. Perhaps the public would be "over-informed."

[xxviii] Shades of FEMA! (Hurricanes Katrina – 2005 and Superstorm Sandy – 2012)

[xxix] "…Those who cannot remember the past are condemned to repeat it…" George Santayana